Reviews and critical articles covering the entire field of normal anatomy (cytology, histology, cyto- and histochemistry, electron microscopy, macroscopy, experimental morphology and embryology and comparative anatomy) are published in Advances in Anatomy, Embryology and Cell Biology. Papers dealing with anthropology and clinical morphology that aim to encourage cooperation between anatomy and related disciplines will also be accepted. Papers are normally commissioned. Originalpapers and communications may be submitted and will be considered for publication provided they meet the requirements of a review article and thus fit into the scope of "Advances". English language is preferred.

It is a fundamental condition that submitted manuscripts have not been and willnot simultaneously be submitted or published elsewhere. With the acceptance of a manuscript for publication, the publisher acquires full and exclusive copyright for all languages and countries.

Twenty-five copies of each paper are supplied free of charg

Manuscripts should be addressed to

Co-ordinating Editor

Prof. Dr. H.-W. KORF , Zentrum der Morphologie, Universität Frankfurt, Theodor-Stern Kai 7,
60595 Frankfurt/Main, Germany
e-mail: korf@em.uni-frankfurt.de

Editors

Prof. Dr. F. BECK, Howard Florey Institute, University of Melbourne, Parkville, 3000 Melbourne, Victoria, Australia
e-mail: fb22@le.ac.uk

Prof. Dr. F. CLASCÁ , Department of Anatomy, Histology and Neurobiology
Universidad Autónoma de Madrid, Ave. Arzobispo Morcillo s/n, 28029 Madrid, Spain
e-mail: francisco.clasca@uam.es

Prof. Dr. D.E. HAINES, Ph.D., Department of Anatomy, The University of Mississippi Med. Ctr.,
2500 North State Street, Jackson, MS 39216–4505, USA
e-mail: dhaines@anatomy.umsmed.edu

Prof. Dr. N. HIROKAWA, Department of Cell Biology and Anatomy, University of Tokyo,
Hongo 7–3–1, 113-0033 Tokyo, Japan
e-mail: hirokawa@m.u-tokyo.ac.jp

Dr. Z. KMIEC, Department of Histology and Immunology, Medical University of Gdansk,
Debinki 1, 80-211 Gdansk, Poland
e-mail: zkmiec@amg.gda.pl

Prof. Dr. R. PUTZ, Anatomische Anstalt der Universität München,
Lehrstuhl Anatomie I, Pettenkoferstr. 11, 80336 München, Germany
e-mail: reinhard.putz@med.uni-muenchen.de

Prof. Dr. J.-P. TIMMERMANS, Department of Veterinary Sciences, University of Antwerpen,
Groenenborgerlaan 171, 2020 Antwerpen, Belgium
e-mail: jean-pierre.timmermans@ua.ac.be

212
Advances in Anatomy, Embryology and Cell Biology

Co-ordinating Editor

H.-W. Korf, Frankfurt

Editors

F.F. Beck · F. Clascá · D.E. Haines · N. Hirokawa
Z. Kmiec · R. Putz · J.-P. Timmermans

For further volumes:
http://www.Springer.com/series/102

Andrzej T. Slominski,
Michal A. Zmijewski,
Cezary Skobowiat, Blazej Zbytek,
Radomir M. Slominski,
Jeffery D. Steketee

Sensing the Environment: Regulation of Local and Global Homeostasis by the Skin's Neuroendocrine System

With 23 figures

 Springer

Andrzej T. Slominski
University of Tennessee
  Health Science Center
Department of Pathology and Laboratory
Medicine, and Department of Medicine
Memphis, TN, USA
aslomins@uthsc.edu

Michal A. Zmijewski
Medical University of Gdansk
Department of Histology
Gdansk, Poland

Cezary Skobowiat
University of Tennessee
  Health Sciences Center
Department of Pathology and Laboratory
Medicine
Memphis, TN, USA

Blazej Zbytek
University of Tennessee Health
  Science Center
Department of Pathology and Laboratory
Medicine
Memphis, TN, USA

Radomir M. Slominski
University of Tennessee Health
  Science Center
Memphis, TN, USA

and

Jagiellonian University Medical College
Krakow, Poland

Jeffery D. Steketee
University of Tennessee Health
  Science Center
Department of Pharmacology
Memphis, TN, USA

ISSN 0301-5556
ISBN: 978-3-642-19682-9   ISBN: 978-3-642-19683-6 (eBook)
DOI: 10.1007/978-3-642-19683-6
Springer Heidelberg New York Dordrecht London

Library of Congress Control Number: 2012939853

Abstract

Skin, the body's largest organ, is strategically located at the interface with the external environment where it detects, integrates, and responds to a diverse range of stressors including solar radiation. It has already been established that the skin is an important peripheral neuro-endocrine-immune organ that is tightly networked to central regulatory systems. These capabilities contribute to the maintenance of peripheral homeostasis. Specifically, epidermal and dermal cells produce and respond to classical stress neurotransmitters, neuropeptides, and hormones. Such production is stimulated by ultraviolet radiation (UVR), biological factors (infectious and noninfectious), and other physical and chemical agents. Examples of local biologically active products are cytokines, biogenic amines (catecholamines, histamine, serotonin, and N-acetyl-serotonin), melatonin, acetylocholine, neuropeptides including pituitary (proopiomelanocortin-derived ACTH, β-endorphin or MSH peptides, thyroid-stimulating hormone) and hypothalamic (corticotropin-releasing factor and related urocortins, thyroid-releasing hormone) hormones as well as enkephalins and dynorphins, thyroid hormones, steroids (glucocorticoids, mineralocorticoids, sex hormones, 7-δ steroids), secosteroids, opioids, and endocannabinoids. The production of these molecules is hierarchical, organized along the algorithms of classical neuroendocrine axes such as hypothalamic–pituitary–adrenal axis (HPA), hypothalamic–thyroid axis (HPT), serotoninergic, melatoninergic, catecholaminergic, cholinergic, steroid/secosteroidogenic, opioid, and endocannbinoid systems. Dysregulation of these axes or of communication between them may lead to skin and/or systemic diseases. These local neuroendocrine networks are also addressed at restricting maximally the effect of noxious environmental agents to preserve local and consequently global homeostasis. Moreover, the skin-derived factors/systems can also activate cutaneous nerve endings to alert the brain on changes in the epidermal or dermal environments, or alternatively to activate other coordinating centers by direct (spinal cord) neurotransmission without brain involvement. Furthermore, rapid and reciprocal communications between epidermal and dermal and adnexal compartments are also mediated by neurotransmission including antidromic modes of conduction. In conclusion, skin cells and skin as an organ coordinate and/or regulate not only peripheral but also global homeostasis.

Acknowledgments

The projects described were supported by grants R01AR052190, R01AR047079, and 1R01AR056666-01A2 from the NIH/NAIMS and grants IBN-9405242, IBN-9604364, and IOS-0918934 from the NSF to ATS and grants from the Polish Ministry of Science and Higher Education N405 623238 and N402 662840 to MAZ. The chapter on melatonin is dedicated to Dr. Aaron B. Lerner who has discovered and defined the structure of melatonin. The chapter on CRF and CRF receptors is dedicated to Dr. Wylie Vale who codiscovered the CRF. We are grateful to Professor Zbigniew Kmiec for his critical comments, intellectual input, careful editorial work, and time committed to this manuscript.

Contents

List of Abbreviations

11β-HSD	11β-Hydroxysteroid dehydrogenase
2-AG	2-Arachidonoylglycerol
5-HT	5-Hydroxytryptamine, serotonin
5-HT$_{1-7}$	5-HT receptors types 1–7
5-HTOL	5-Hydroxytryptophanol
5HTT	Serotonin transporter
5MTOL	5-Methoxytryptophol
5MTT	5-Methoxytryptamine
6BH4	6R-L-erythro-5, 6,7,8-tetrahydrobiopterin
7DHC	7-Dehydrocholesterol
7DHP	7-Dehydropregnenelone
7TM	7 Transmembrane
AAD	L-Amino acid decarboxylase
AANAT	Arylalkylamine N-acetyltransferase
ACTH	Adrenocorticotropic hormone
AEA	N-arachidonoylethanolamide, anandamide
AFMK	N1-acetyl-N2-formyl-5-methoxykynuramine
AMK	N1-acetyl-5-methoxykynuramine
AVP	Arginine vasopressin
B	Corticosterone
CaR	Calcium receptor
CB1/2	Endocannabinoid receptor 1/2
CGRP	Calcitonin gene-related peptide
cis/trans-UCA	cis/trans-Urocanic acid
CNS	Central nervous system
COMT	Catechol-methyl transferase
COR	Cortisol
CORT	Corticosterone
CRE	cAMP response element
CREB	cAMP response element-binding
CRF/CRH	Corticotropin-releasing factor/hormone
CRF1or 2	Corticotropin-releasing factor receptor 1 or 2
CYP	Cytochrome P

DAG	Diacylglycerol
DHEA	Dehydroepiandrosterone
DHT	Dihydrotestosterone
DOC	Deoxycorticosterone
DOR	Delta opioid receptor
DRG	Dorsal root ganglia
DYN A	Dynorphin A
ECD	Extracellular domain
ECS	Endocannabinoids
END	Endorphins
ENK	Enkephalins
ERK	Extracellular signal-related kinase
F	Cortisol
FAAH	Fatty acidamid hydrolase
GABA	Gamma-aminobutyric acid
GI	Gastrointestinal tract
GIRK	G-protein-regulated inwardly rectifying potassium channels
GM-CSF	Granulocyte-macrophage colony-stimulating factor
GPCRs	G-protein-coupled receptors
HaCaT	Human immortalized keratinocytes
HIOMT	Hydroxyindole-O-methyltransferase
HPA axis	Hypothalamic–pituitary–adrenal axis
HPT axis	Hypothalamic–pituitary–thyroid axis
HSD	Hydroxysteroid dehydrogenase
IL	Interleukin
IP3	Inositol trisphosphate
KOR	Kappa-opioid receptor
L-DOPA/DOPA	L-3,4-Dihydroxyphenylalanine
LENK	Leu-Enkephalin
MAO	Monoamine oxidase
MAPK	Mitogen-activated protein kinases
MC2-R	Melanocortin receptor type 2
MEK	Mitogen-activated protein and extracellular signal-regulated kinase
MENK	Met-Enkephalin
MOR	Mu-opioid receptor
MSH	Melanocyte-stimulating hormone
MT1 or 2	Melatonin receptor type 1 or 2
NAS	N-acetylserotonin
NAT	Arylamine N-acetyltransferase
NF-κB	Nuclear factor kappa-light-chain-enhancer of activated B cells
NGF	Nerve growth factor
NO	Nitric oxide

NOS	Nitric oxide synthase
NQO1 or 2	Guinone oxidoreductase 1 or 2
OR	Opioid receptor
P450scc	Cytochrome P450 side-chain cleavage enzyme
PDYN	Prodynorphin
PEA	N-palmitoylethanolamide
PENK	Proenkephalin
PH	Phenylalanine hydroxylase
PKA	Protein kinase A
PLC	Phospholipase C
POMC	Proopiomelanocortin
PVN	Paraventricular nucleus
ROS	Radical oxygen species
SP	Substance P
Src	v-src sarcoma (Schmidt-Ruppin A-2) viral oncogene homolog (avian)
SRC	Steroid receptor coactivator
T3	Triiodothyronine
T4	Thyroxine
TGFβ-2	Transforming growth factor beta-2
TH	Tyrosine hydroxylase
THC	Tetrahydrocannabinol
TPH	Tryptophan hydroxylase
TRH	Thyroid-releasing hormone
TRH-R	TRH receptor
TRpOH	5-Hydroxytryptophan
TRPV	Transient receptor potential vanilloid
TRα or β	Thyroid hormone receptor α or β
TSH	Thyroid-stimulating hormone
Tyr	Tyrosinase
URC	Urocortin
Urc 1-3	Urocortin types 1-3
UV	Ultraviolet
UVA	Ultraviolet A radiation
UVB	Ultraviolet B radiation
UVR	Ultraviolet radiation
VDR	Vitamin D3 receptor

Chapter 1
Introduction

1.1 General Overview

The strategic location of the skin as the barrier between the environment and internal milieu determines its critical function in the preservation of body homeostasis, and ultimately organism survival (Slominski 2005; Slominski and Wortsman 2000; Slominski et al. 2000c; Zmijewski and Slominski 2011). It also exposes skin to numerous pathological agents, processes, and events. Thus, the capability to locally recognize, discriminate, and integrate various signals within a highly heterogeneous environment, and to immediately launch appropriate responses, is a vital property of skin (Slominski and Wortsman 2000). These skin functions are integrated into the skin immune, pigmentary, epidermal, and adnexal systems, and are in continuous communication with the systemic immune, neural, and endocrine systems (Arck et al. 2006; Slominski 2009c; Slominski and Wortsman 2000; Slominski et al. 2004c, 2007a; Stenn and Paus 2001).

These fundamental functions result from the location of the skin, which is the largest body organ, at the interphase between external and internal environment, requiring development of efficient sensory and effector capabilities to differentially react to changes in external environment. They are represented by inducible production of biologically active compounds (hormones, neurohormones, and neurotransmitters) that act both locally and at the systemic levels (Fig. 1.1).

The skin being continuously exposed to many external biological or environmental factors (acute transfers of solar, thermal, or chemical energy) had to evolve optimal mechanism(s) to protect, restore, or maintain local and global homeostasis in relation to hostile environment (Slominski et al. 1993b, 2000c; Slominski and Pawelek 1998; Slominski and Wortsman 2000). We have proposed that precise coordination and execution of these responses are mediated by a cutaneous neuroendocrine system, which also is able to reset the body homeostatic adaptation mechanisms (Slominski and Wortsman 2000). Superimposed on this is the impact of psychological stress on skin physiology and pathology,

A.T. Slominski et al., *Sensing the Environment: Regulation of Local and Global Homeostasis by the Skin's Neuroendocrine System*, Advances in Anatomy, Embryology and Cell Biology 212, DOI 10.1007/978-3-642-19683-6_1, © Springer-Verlag Berlin Heidelberg 2012

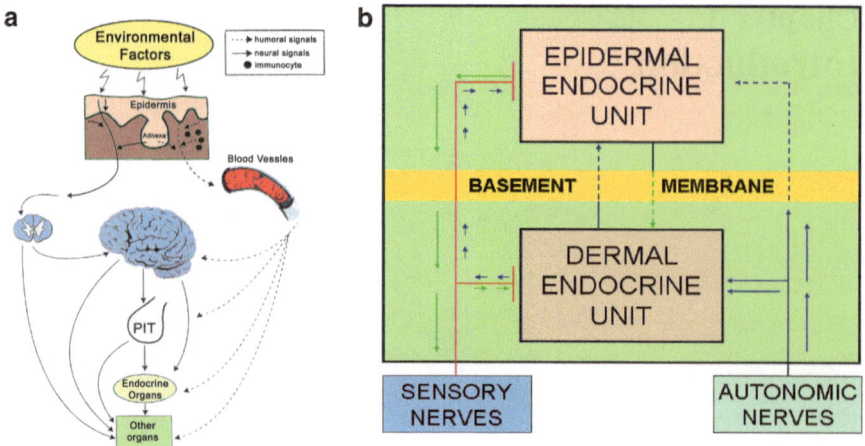

Fig. 1.1 Skin senses changes in the environment through cutaneous neuroendocrine system, which computes and translates the received information into chemical, physical, and biological messengers that regulate global (A and B) and local (B) homeostasis. These signals travel via through humoral, immune, or neural pathways to reach the central nervous, endocrine, and immune systems as well as other organs. Reproduced with permission from the Endocrine Society (Slominski and Wortsman 2000)

placed in the context of the bidirectional brain–skin communication (Arck et al. 2006; Slominski 2005; Slominski et al. 2008b). To summarize, in reaction to changing external and also internal environment, the skin can generate signals to produce rapid (neural) or slow (humoral or immune) responses at the local and systemic levels (Fig. 1.1).

Coordination between these local and systemic responses is mediated by the skin neuroendocrine system (Slominski and Wortsman 2000), which employs local equivalents of the hypothalamo–pituitary–adrenal axis (HPA) (Slominski et al. 2007a), hypothalamo–pituitary–thyroid (HPT) axis (Pisarchik and Slominski 2002; van Beek et al. 2008), catecholaminergic (Schallreuter et al. 1997), serotoninergic, melatoninergic (Slominski et al. 2005c, 2008a), cholinergic (Grando 2006; Grando et al. 2006), steroidogenic (Slominski et al. 2008b), and secosteroidogenic (Bikle 2010; Holick 2003; Slominski et al. 2010) systems (Fig. 1.2). Given their common embryonic origins, it is not surprising that skin shares numerous mediators with the CNS and endocrine system. Recent research has revealed that skin also harbors complex opioidogenic (Grando et al. 1995; Slominski et al. 2011c) and cannabinoidogenic (Biro et al. 2009) systems, the role of which in the maintenance of cutaneous homeostasis is currently being intensively explored.

In this monograph, we will discuss the role of various components of the skin neuroendocrine system in sensing the environment with subsequent regulation of local and global homeostasis with a main focus on the algorithms of classical neuroendocrine axes.

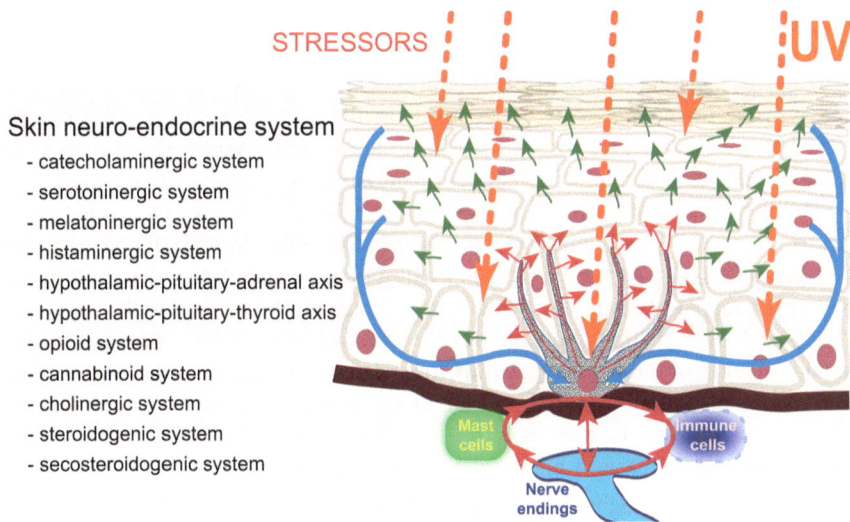

Fig. 1.2 Skin neuroendocrine system follows the algorithms of classical neuroendocrine or endocrine systems. It also forms a natural platform of signal exchange between internal organs and environment. For this purpose, skin cells not only are subjected to neurohormonal regulation but also do produce neuropeptides, biogenic amines, melatonin, opioids, cannabinoids, acetylcholine, steroids, secosteroids, as well as growth factors and cytokines. Skin neuroendocrine system also entrains immune cells to act as cellular messengers at distal sites

1.2   An Overview of Histology and Anatomy

Since histology and anatomy of the skin has been extensively reviewed in three major dermatology and dermatopathology textbooks (Bolognia et al. 2008; Fitzpatrick et al. 1993; Weedon 2010), below we provide only a short overview. The most external layer of the skin, the epidermis, is derived from the ectoderm, and is characterized by a constant renewal. The main constituents of the epidermis, keratinocytes, are either self-replicating in the basal layer (about 50% of basal layer keratinocytes are in this state) or differentiating toward the surface (another 50%). The whole process of differentiation lasts about 31 days. The keratinocytes of succeeding layers (spinous and granular) gradually flatten to form a solid cornified layer that is subsequently shed (this takes another 14 days on average). The intermediate filaments, cytokeratins, are the most important structural elements of the keratinocyte. In the epidermis, cytokeratins 5 and 14 are main cytokeratins in basal keratinocytes and cytokeratins 1 and 10 in differentiating ones. The cornified layer is formed by various cross-linked proteins and lipids. Apart from keratinocytes, there are other cells in the epidermis whose function is more regulatory than structural. Examples are melanocytes, derived from neural crest, which reside in the basal layer. Their density varies in different parts of the body from 1 in 4 to 10 basal keratinocytes. Melanin, protective pigment produced by these cells, is transferred from melanocytes through their processes to approximately neighboring

36 keratinocytes (to form epidermal-melanin unit) by the process of apocopation. Melanin not only absorbs UV radiation but also serves as a scavenger of reactive oxygen species and miscellaneous chemical compounds. The Langerhans cells are derived from the bone marrow. They reside at different levels of the epidermis and engulf foreign antigens. They transport them to the lymph nodes and present in the context of MHC antigens to T lymphocytes initiating the adaptive immune response.

The dermis is derived from mesoderm. Its bulk is composed of collagen and elastic fibers and glycosaminoglycans. The main collagen of reticular dermis is collagen type I. Collagen type III is present in the adventitial dermis (papillary and peri-appendageal). Elastic fibers are arranged in a parallel manner in the superficial dermis including elaunin fibers (made of microfibrils with elastin core) and perpendicular manner in the papillary dermis (oxytalan fibers made of microfibrils only). Collagen gives skin its strength, elastic fibers its elasticity (ability to retract), and glycosaminoglycans its substance. Various inflammatory cells typically reside in the dermis and increase in numbers when need arises. Dermal vasculature forms superficial and deep dermal plexuses that are connected by straight collaterals. Superficial plexus sends papillary loops toward the surface. Of note, the epidermis does not have its own vasculature and is being nourished through exchange of substances provided by the most superficial parts of papillary capillaries. Glomus bodies (Sucquet-Hoyer canals) are important for local thermoregulation.

Skin appendages are of epidermal origin. The hairs cover most of the body. Terminal and vellus hairs differ in their size and function. The hairs undergo cyclic changes of growth (anagen, about 90% of scalp hair, lasts 3–10 years), involution (catagen, 1%, lasts weeks), and rest (telogen, 10%, lasts few months). Of note, different hairs on the body, even directly neighboring, are in different phases of the growth cycle. This is a major difference between humans and animals that shed hair cyclically. The sebaceous glands are usually associated with hair and secrete protective lipid substances by a holocrine mechanism. The coiled eccrine glands are located in the subcutis; their straight ducts transverse the dermis and end in coiled fashion in the acrosyringia of the epidermis. The primary sweat is hyper- or isotonic and becomes hypotonic during passage through the excretory ducts. Sweat production is the most important thermoregulatory mechanism in humans. Apocrine glands are distributed only in some areas of the body (axillae, genital, ear, and eyelid) and have probably only vestigial function in humans.

Last, but not least, the subcutaneous fat tissue is a third important layer of the skin. Fat lobules forming it are separated by fibrous septae transverse rich in vasculature. The adipose tissue is mostly of white type and has important function in isolation, cushion, and energy storage. Often quoted to be body's largest immune/endocrine organ (about 15% of body weight and average surface of about 2 m^2), skin is a source of multiple mediators and cytokines that act not only locally but also systemically. On the other hand, components of skin respond to internal stimuli and mediators preserving body homeostasis and appropriate functioning.

Skin is studied by a variety of methods. The classic histological slides, prepared from formalin-fixed tissue and stained with hematoxylin and eosin paired with various special stains and by immunohistochemical methods, are the tools of both dermatopathologist and researcher. Direct immunofluorescence is a complementary method used for both diagnosis and research. Frozen sections are stained here with antibodies against immunoglobulins, complement, and fibrinogen. Different patterns are observed and yield diagnostic information. Popular research tools are the ex vivo skin cell cultures. Both primary (with definite number of cell divisions) and continuous (indefinite number of cell divisions) cell cultures are being used. To better model the conditions present at the skin as tissue, the ex vivo organ cultures are also used. A plethora of cell and molecular biology methods have been applied for studies of both cell and organ cultures. Some popular examples are Western blot, PCR, confocal microscopy, and gene microarrays.

1.3   An Overview of Skin Innervation

The skin extensive neural network represented by somatosensory and autonomic nerve fibers has been described in detail in several reviews and books (Bolognia et al. 2008; Fitzpatrick et al. 1993; Roosterman et al. 2006; Siemionow et al. 2011; Slominski and Wortsman 2000; Weedon 2010; Yosipovitch 2010). Therefore, below we provide only a short overview.

In the skin, receptors localized on the primary afferent nerve terminals transduce various sensory stimuli, generated upon changes in temperature, pH, and the presence of inflammatory mediators, and convey them to the specific areas of the CNS what results in the perception of pain, itching neuroinflammation, as well as somatic responses of other organs and tissues. The perikarya of cutaneous sensory fibers are localized either in the dorsal root ganglia (DRG) or, those innervating the face and upper neck, in the trigeminal ganglion. Both unmyelinated (C) and myelinated (A) fibers of unipolar sensory cells conduct thresholds at 0.5–2 m/s and 4–70 m/s, respectively. The ortho- and antidromic conduction of afferent nerve fibers results in simultaneous signal transduction and release of neurotransmitters (mainly substance P and CGRP) at the same site. The sensory axons make synapses in dorsal spinal cord neurons depending on somatotopic map of the part of the body surface innervated by the relevant spinal segments. The major ascending routes for sensory cutaneous inputs are via the dorsal column nucleus (DCN) or lateral cervical nucleus (LCN). Both of them transmit to the thalamus, which is a coordinator station for sensory imputes receiving and sending neural signals to somatosensory cortex, midbrain, and hypothalamus—the headquarters of the autonomic nervous system. The connection between thalamus and hypothalamic paraventricular nuclei constitutes important element joining cutaneous stimuli with centers which control body homeostasis and endocrine system, including HPA axis. Also, cutaneous afferent stimuli from face run in the trigeminal root and upon switch in trigeminal nucleus terminate in the thalamus (Siemionow et al. 2011).

The cutaneous innervation has traditionally been considered to consist of a plexus of fibers in the reticular layer of dermis and a more superficial plexus in the papillary layer, with the majority of sensory endings located in the subpapillary dermis. Recent advances in immunohistochemistry provided an evidence for the existence of intraepidermal nerve fibers (reviewed in Bolognia et al. 2008; Legat and Wolf 2009; Roosterman et al. 2006; Slominski and Wortsman 2000; Waller et al. 2011). Intraepidermal nerve terminals associated with Merkel cells, cold receptors, and high-threshold mechanoreceptors have been identified in the basal layer of the epidermis. Thin nerve fibers travel through the dermis, extend into epidermis, and terminate with or without branching in all layers of epidermis including stratum corneum. Waller et al. 2011). The density of epidermal nerve fibers changes during aging and in many pathological conditions like diabetes, psoriasis, or upon ultraviolet radiation. Therefore, quantification of the epidermal nerve fibers' density was proposed to be a valuable prognostic marker for the evaluation of the disease progress (Fromy et al. 2010; Legat and Wolf 2009; Roosterman et al. 2006; Waller et al. 2011).

In the skin, cutaneous nerve fibers have principally sensory character, with an additional component of autonomic nerve fibers distributed exclusively in dermis. Most of them are found in the mid-dermis and the papillary dermis. The autonomic nerves supply arterioles, glomus bodies, hair erector muscles, and apocrine and eccrine glands. A rich network of autonomic and sensory nerve fibers surrounds especially hair follicles, pilosebaceous units, and eccrine and apocrine glands. The sensory and autonomic networks show regional differences according to anatomic location and also have topographical specificity by distributing into well-defined areas called dermatomes. The autonomic nerve fibers in the skin predominantly derive from sympathetic (cholinergic, catecholaminergic, and non-adrenergic/non-cholinergic) and, in the face, rarely parasympathetic (cholinergic) neurons.

In addition to the classic neurotransmitters like acetylcholine, noradrenaline, and serotonin, the postganglionic autonomic nerves in the skin predominantly release also neuropeptides (neuropeptide Y, galanin, vasoactive intestinal peptide, and β-endorphin) and biologically active substances (nitric oxide, ECS) which act as co-transmitters. These compounds modulate the release and activity of the main neurotransmitters and also directly affect targeted cells. Neuropeptides released from cutaneous nerves via a paracrine, juxtacrine, or endocrine manner act on target cells which express specific receptors that are appropriately coupled to an intracellular signal transduction pathway or ion channels, which, when activated, may result in the activation of biological responses such as erythema, edema, hyperthermia, and pruritus.

Chapter 2
Biogenic Amines in the Skin

2.1 An Overview

It has been documented that skin resident cells can produce and further metabolize catecholamines, serotonin, and histamine (Fitzpatrick et al. 1993; Gillbro et al. 2004; Schallreuter et al. 1995; Slominski et al. 2005c). These biogenic amines not only regulate phenotype of skin cells cultured in vitro but also can affect skin functions and may have systemic effects (Schallreuter et al. 1997; Slominski and Wortsman 2000; Slominski et al. 2005c). The functional activity of biogenic amines in the skin is mediated through the interactions with specific cell surface receptors (Gillbro et al. 2004; Nordlind et al. 2008; Slominski et al. 2003d); however, non-receptor effects are also considered.

2.2 Catecholamines

2.2.1 Production and Metabolism

Nonessential aromatic amino acid L-tyrosine, depending on the cell type and enzymatic context, serves as a direct precursor to catecholamines, tyramine/octopamine (Yen 2001), and melanin pigment (Slominski et al. 2004c). To serve these diverse functions, L-tyrosine is either delivered through the gastrointestinal tract (GI) or produced through phenylalanine hydroxylase (PH)-mediated hydroxylation of L-phenylalanine (Blau et al. 2010; Schallreuter et al. 2008b). L-tyrosine is hydroxylated to L-dihydroxyphenylalanine (L-DOPA) by either tyrosine hydroxylase (TH) or tyrosinase (Tyr), or decarboxylated to tyramine by L-amino acid decarboxylase (AAD) (Fig. 2.1). L-DOPA is further decarboxylated to dopamine by AAD with subsequent hydroxylation and methylation reactions to generate norepinephrine or epinephrine, all of them being oxidized by monoamine oxidase (MAO) or methylated by catechol-methyl transferase (COMT) (Fig. 2.1). L-DOPA

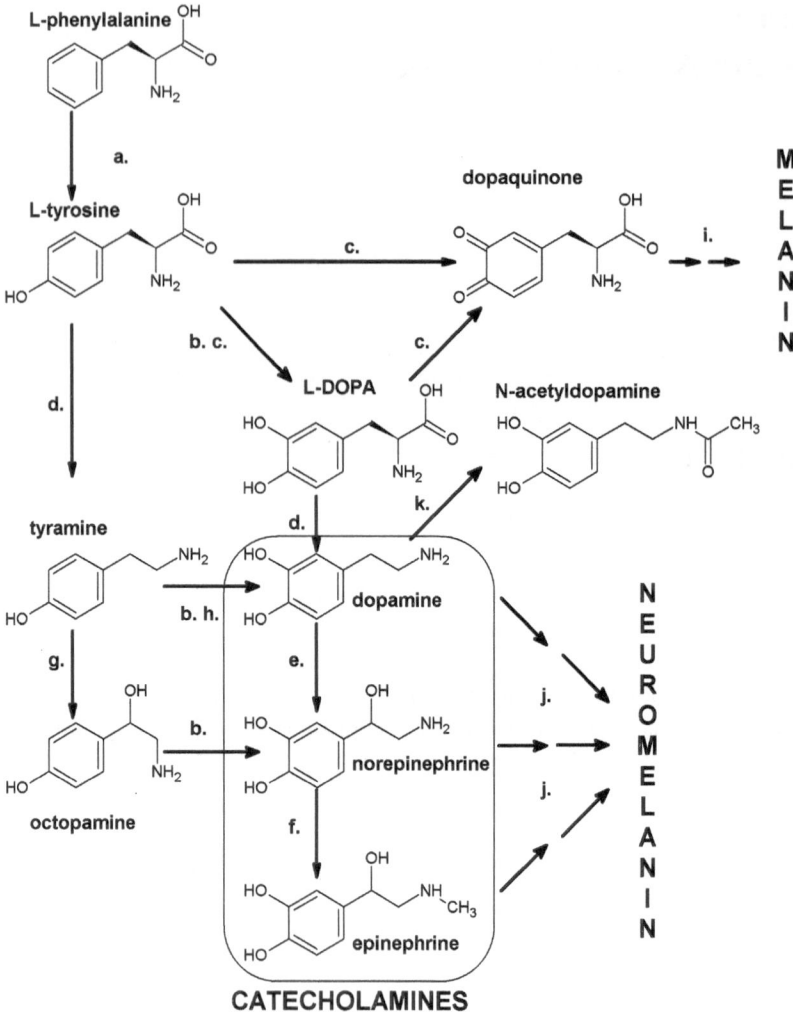

Fig. 2.1 Catecholamine synthesis in the skin. The common pathway in the skin requires its consecutive hydroxylations of L-phenylalanine [mediated by phenylalanine hydroxylase (PH)] to L-tyrosine with following hydroxylation by tyrosine hydroxylase (TH) or tyrosinase to produce L-dihydroxyphenylalanine (L-DOPA). L-DOPA is either oxidized to DOPA quinone with following multistep transformation to melanin or serves as a substrate for synthesis of catecholamines. The skin expresses complete enzymatic machinery required for dopamine synthesis (L-amino acid

and catecholamines can also be oxidized by either tyrosinase or metal cations to form melanin and neuromelanin pigments (Fitzsimons et al. 2002; Lassalle et al. 2003; Park et al. 2009; Slominski et al. 2004c) (Fig. 2.1).

Human epidermal keratinocytes and melanocytes have the capability to synthesize the catecholamines from L-tyrosine with sequential production of L-DOPA, dopamine, norepinephrine, and epinephrine through the action of classical enzymes listed above with the subsequent inactivation of catecholamines by MAO or COMT (Fig. 2.2) (Fuziwara et al. 2005; Gillbro et al. 2004; Schallreuter et al. 1992, 1995). Interestingly, acetylation of dopamine to N-acetylDOPA has also been described in the hamster skin (Gaudet et al. 1993). Activity of TH and PH depends on local availability of their essential cofactor/electron donor, i.e., 6R-L-erythro-5, 6, 7, 8-tetrahydrobiopterin (6BH4) as demonstrated for the first time by Schallreuter's group (Schallreuter et al. 1994, 1997). Importantly, Schallreuter and coworkers demonstrated de novo synthesis/recycling/regulation of 6BH4 in both human epidermal keratinocytes and melanocytes as well as in hair follicles (Schallreuter et al. 1997, 1998). Furthermore, AAD activity requires pyridoxal phosphate (PP) as the cofactor, the cutaneous availability of which is regulated locally (Coburn et al. 2003). Lymphocytes and other immune cells can also represent an additional source of catecholamines: L-DOPA production with its further transformation to epinephrine and norepinephrine has been shown in human lymphocytes (Musso et al. 1997) as well as in Langerhans cells (Falck et al. 2004). An additional cutaneous source of catecholamines is their dermal release from adrenergic nerve fibers (Fitzpatrick et al. 1993). A challenging task in current skin biology is to determine which skin cells and adnexal structures have similar capability of de novo synthesis of catecholamines and what is the final product in different compartments.

An important alternative source of L-DOPA for cutaneous catecholamines is its production via the tyrosine hydroxylase activity of tyrosinase that, depending on the intracellular environment including acidic pH, may not undergo oxidation but will diffuse or be transported to other cells or systemic circulation (Slominski et al. 2004c, 2011a). In fact, diffuse "melanocytic organ" can provide DOPA or its adducts to systemic circulation to serve either as a precursor for further modifications or as a bioregulator (Slominski et al. 1993a, 2011a; Zmijewski and Slominski 2009a). A role for tyrosinase-derived L-DOPA is supported by findings that retinal network adaptation to bright light requires tyrosinase-dependent production of DOPA (Page-McCaw et al. 2004). This phenomenon represents the TH-independent pathway of peripheral dopamine synthesis (Eisenhofer et al. 2003) and

---

Fig. 2.1 (continued) decarboxylase, AAD) and its subsequent conversion into norepinephrine (dopamine β-hydroxylase) and methylation (phenylethanolamine N-methyltransferase) to form epinephrine. An alternative pathway of catecholamine synthesis involves decarboxylation of L-tyrosine to tyramine, which in turn is hydroxylated by TH (and Cyp2D) or dopamine β-hydoxylase to octopamine or dopamine, respectively. Octopamine could be metabolized to norepinephrine by TH. This alternative pathway which is present in invertebrates remains to be tested in the skin. Catecholamines also undergo oxidation to corresponding quinoinones with further multistep transformation to neuromelanin, a process similar to melanogenesis starting from L-DOPA

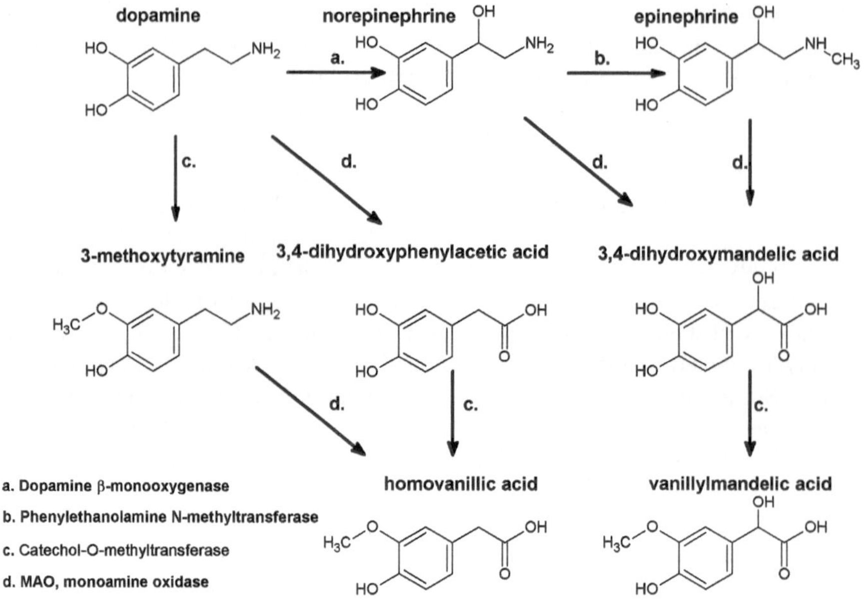

Fig. 2.2 Catecholamine catabolism. Catecholamines are deactivated by L-monoamine oxidase (MAO) and Catechol-*O*-methyltransferase (COMT) leading to the synthesis of homovanillic acid (from dopamine) or vanillylmandelic acid from norepinephrine or epinephrine. Alternatively, as shown for dopamine metabolism the order of reaction may be changes with COMT acting first and then followed by MAO

it can regulate activities of melanocytes and immune cells (Slominski and Paus 1990; Slominski et al. 1998c). These findings are in agreement with our hypothesis that L-tyrosine and L-DOPA can have hormone- and neurotransmitter-like roles (Slominski and Paus 1990, 1994; Slominski et al. 2011a), with melanocytes acting as important regulators of catecholamines' availability in the skin (Slominski et al. 1993a).

2.2.2 *Bioregulatory Role of Catecholamines in the Skin*

2.2.2.1 Dopamine Receptors

There are five subtypes of dopamine receptors, and they have been categorized into two families, i.e., D1-like receptors (D1 and D5) and D2-like receptors (D2, D3, and D4) (Watson 1994). The D1-like receptor agonists stimulate Gs-dependent intracellular production of cAMP (Missale et al. 1998). The D2-like receptor agonists activate Gi proteins and inhibit intracellular cAMP signaling pathway (Missale et al. 1998; Watson 1994). In addition, via Gβγ subunits, D2-like receptors are capable of inhibiting N- and L-type calcium channels which results in the activation of G-protein-regulated inwardly rectifying potassium channels (GIRKs) (Beaulieu and Gainetdinov 2011). After D2-like receptors were identified in the keratinocytes (Fuziwara et al. 2005) they were found to play a significant role in the maintenance of epidermal barrier homeostasis. Application of D2-like receptor agonists accelerated barrier recovery, whereas D2-like receptor antagonists delayed it. Actual receptor subtypes localize to different parts of the epidermis: D4 is localized in the uppermost part of the epidermis and D2 is localized in the basal layer of the epidermis where it plays a role in the regulation of cell proliferation (Fuziwara et al. 2005). It remains to be tested whether dopamine is also regulating epidermal and follicular pigmentary systems as well as adnexal functions including hair follicle.

Dopamine receptors on lymphocytes exert differential effects. Dopaminergic signaling through D2-like receptors of T lymphocytes showed an immunostimulatory effect (Besser et al. 2005), whereas signaling through D1-like receptors had immunoinhibitory effect (Saha et al. 2001). Dopamine also inhibits proliferation of human lymphocytes and causes apoptosis of peripheral blood mononuclear cells (Bergquist et al. 1997). IL-6 (and other cytokines) stimulates a development of a subtype of T lymphocytes capable of producing IL-17 (and other cytokines), i.e., Th17 lymphocytes. Th17 lymphocytes constitute relatively recently described branch of immune responses (Harrington et al. 2006). Dopamine released by dendritic cells induces IL-6–Th17 axis and upregulates synovial inflammation (Nakano et al. 2011). The IL-6–Th17 axis plays a role in the pathogenesis of inflammatory diseases including rheumatoid arthritis. It can therefore be deduced that dopamine may also have various differential modulatory roles in the skin immune system.

2.2.2.2 Adrenergic Receptors

The adrenergic receptors belong to the classic seven-transmembrane G-protein-coupled receptor (GPCR) family. These receptors respond to catecholamines and can be subdivided into subtypes of α and β families, based on their differential pharmacological responses and protein sequences (Lands et al. 1967). More specifically, these receptors are defined, in part, by their endogenous ligand affinity to β receptors having a higher affinity to epinephrine when compared to norepinephrine, and to α receptors having a higher affinity for norepinephrine. Alpha adrenergic receptors can be further subdivided into $\alpha 1$ and $\alpha 2$, and β receptors can be further subdivided into $\beta 1$, $\beta 2$, and $\beta 3$ subtypes. The $\alpha 1$ ($\alpha 1a$, $\alpha 1b$, and $\alpha 1d$) receptors couple to phospholipase C via $Gq\alpha$ and stimulate the formation of diacylglycerol and inositol trisphosphate (Cotecchia 2010). The $\alpha 2$ ($\alpha 2a$, $\alpha 2b$, and $\alpha 2c$) receptors couple to $Gi\alpha$ and inhibit the formation of cAMP, whereas β receptors are positively coupled to the formation of cAMP via $Gs\alpha$ (Hein 2006).

Various receptors of both α and β subfamilies of adrenergic receptors are present on epidermal and dermal cells (Grando et al. 2006; Schallreuter et al. 1995). As expected, α and β receptors are also expressed in dermal blood vessels. Their activation by catecholamines causes vasoconstriction and decreases vascular permeability (Ding et al. 1995; Harada et al. 1996).

Keratinocytes express mainly β2 receptors and also α1 receptors (Steinkraus et al. 1992; Drummond et al. 1996; Sivamani et al. 2007). Stimulation of β-adrenergic receptors in epidermal keratinocytes results in increased cAMP production, activation of protein kinase C, and formation of inositol-1,4,5-trisphosphate, calcium influx, and extracellular signal-related kinase (ERK) dephosphorylation through the action of serine/threonine phosphatase PP2A (Chen et al. 2002; Pullar et al. 2001; Schallreuter et al. 1995). Catecholamines stimulate keratinocyte differentiation with increased expression of keratins 1, 10, involucrin, and transglutaminase (Mammone et al. 1998; Schallreuter et al. 1995). Moreover, there is a local gradient of receptor expression with the highest level in basal keratinocytes and decreasing level toward the surface of the epidermis (Schallreuter et al. 1995). This indicates a potential stimulatory functional role of catecholamines in the process of keratinocytes' differentiation. Catecholamine-β2 adrenergic system has been implicated in the pathogenesis of atopic dermatitis, psoriasis, and vitiligo (Sivamani et al. 2007). Expression of β2 receptors is increased in vitiligo and decreased in psoriasis (Schallreuter et al. 1993; Takahashi et al. 1996). In vitiligo, there is an overproduction of 6-BH4 leading to a dysregulation of catecholamine biosynthesis with increased plasma and epidermal norepinephrine levels. This is associated with high numbers of β2 adrenoceptors in differentiating keratinocytes and with a defective calcium uptake in both keratinocytes and melanocytes (Schallreuter et al. 2008a). In atopic eczema, a point mutation in the beta 2-adrenoceptor gene could alter the structure and function of the receptor, thereby leading to a low density of receptors on both keratinocytes and peripheral blood lymphocytes (Schallreuter et al. 1997). It is also known that catecholamines and β receptors have a role in wound healing although their exact role is far from being clarified (Ghoghawala et al. 2008; Pullar et al. 2008) (see also discussion of fibroblast below). The adrenergic beta-receptors not only affect keratinocytes' proliferation and differentiation but also their immune activities. Activation of β receptors on keratinocytes affects expression of β-defensin 3 (Martin-Ezquerra et al. 2011).

Studies on cultured melanoma cell lines have shown that catecholamines can be an additional factor affecting melanogenesis (Howe et al. 1991). Their role in the function of the pigmentary system has been well described in nonhuman systems (reviewed by Slominski et al. 2004c). Human melanocytes express α1 and β2 receptors (Gillbro et al. 2004; Hu 2000; Hu et al. 2000; Scarparo et al. 2000; Schallreuter et al. 1996). Activation of α1 receptors leads to the IP3-DAG signaling (Schallreuter et al. 1996) and β2 receptor activation leads to cAMP signaling (Gillbro et al. 2004). β2 but not α1 receptor activation induces pigmentation (Gillbro et al. 2004; Schallreuter et al. 1996). The expression of β2 receptors on human melanocytes increases in response to UV irradiation (Yang et al. 2006). UVB irradiation increases epinephrine release by cultured keratinocytes that in turn increases pigmentation in co-cultured melanocytes, which is an example of the interactions between these two cell types (Sivamani et al. 2009).

Adrenergic receptors are expressed also on immune cells of the dermis (Steinkraus et al. 1996). Binding of adrenergic agonists to their receptors on lymphocytes has immunostimulatory effect and affects their homing. On the contrary, stimulation of β receptor usually has immunosuppressive effects, but in other model systems can also cause immunostimulation, i.e., increase the number of lymphocytes (Bergmann and Sautner 2002).

Mouse Langerhans cells express α1, β1, and β2 adrenergic receptors (Seiffert et al. 2002), and it was shown that epinephrine and norepinephrine inhibit the ability of Langerhans cells to present antigens (Seiffert et al. 2002).

Agonists of β2 receptors on mast cells inhibit the release of preformed mediators such as histamine, and also newly synthesized mediators such as prostaglandin D2 from mast cells (Okayama and Church 1992). They also inhibit release of inflammatory cytokines from mast cells (Bissonnette and Befus 1997). β receptors are expressed on dermal fibroblasts (Pullar and Isseroff 2006; Pullar et al. 2008). Ligation of β2 receptors activates epidermal growth factor (EGF) receptor and extracellular signal-regulated kinase (ERK) signaling that in turn stimulates fibroblast migration. Binding of agonists to the β2 receptors can also activate protein A kinase (PKA) which can stimulate cell proliferation (Pullar and Isseroff 2006), attenuate collagen gel contraction, and alter actin cytoskeleton and focal adhesion distribution via PKA-dependent mechanisms (Pullar and Isseroff 2006). A link between body stress response system that results in the release of epinephrine and activation of intracellular signaling that leads to DNA damage has been shown recently (Hara et al. 2011). Specifically, in mouse and human fibroblasts binding of agonists to the β2 receptors led to Gs-protein-dependent activation of protein kinase A, followed by the recruitment of beta-arrestins. Then, β-arrestin 1 facilitated AKT-mediated activation of MDM2 and also promoted MDM2 protein binding to and degradation of p53 protein by acting as a molecular scaffold. The degradation of p53 resulted in the lack of protection and DNA damage (Hara et al. 2011).

2.2.2.3 Non-receptor-Mediated Effects of Catecholamines

In the skin there are several potential non-receptor-mediated effects, which are based on autoxidation of catecholamines in alkaline environment with a velocity increased by metal cations (Lassalle et al. 2003; Slominski et al. 2004c). The potential phenotypic implications are predominantly based on the well-documented activity of L-DOPA which through its oxidation products and active melanogenesis can affect functions of immune cells (Slominski and Goodman-Snitkoff 1992; Slominski et al. 2009b). The possible mechanisms of action were discussed previously (Slominski et al. 1998c, 2004c) and, therefore, have been shortly summarized below. L-DOPA dramatically inhibits an in vitro phosphorylation of glycoproteins dependent on the presence of Mn ions indicating action of quinones generated through oxidation of DOPA (Slominski and Friedrich 1992). It can also affect cellular metabolism in melanotic cells (Scislowski et al. 1984, 1985). Also, diffusible products of DOPA oxidation are potent inhibitors of lipid peroxidation

(Memoli et al. 1997), and 5-S-cysteinyldopa inhibits hydroxylation/oxidation reactions induced by the Fenton reaction (Napolitano et al. 1996). The potential cycling from indole to quinone forms of L-DOPA and its derivatives may affect levels of reactive oxygen/nitrogen species or oxidation of intracellular proteins and lipids (Tsang and Chung 2009). Finally, both free and protein-bound L-DOPA can trigger expression of several antioxidant enzymes including superoxide dismutase or NAD(P)H:Quinone oxidoreductase (NQO1) (Nelson et al. 2007). Thus, taking into consideration similar chemical properties of DOPA and catecholamines (products of DOPA enzymatic metabolism), and that their oxidation leads to the production of neuromelanin, one can safely conclude that non-receptor-mediated effects and mechanisms will be similar to that described for DOPA (Slominski et al. 2011a). Taking into consideration the above chemical properties of dopamine, epinephrine, or norepinephrine, one can expect that at micromolar or higher concentrations the predominant effects will be non-receptor-mediated mainly through their oxidation products and neuromelanin polymers generated during this process. It is also possible that some of the phenotypic effects at lower concentrations could also be influenced by oxidative effects.

2.2.2.4 Conclusions

Dopamine, epinephrine, and norepinephrine are produced in the skin resident and nonresident cells. Their phenotypic effects are mediated through activation of dopaminergic and adrenergic receptors, the expression of which is cell-type and cell-context dependent. Their roles in epidermal, dermal, and adnexal as well as skin immune functions remain to be further investigated. There are also non-receptor-mediated mechanisms shared by their precursor, L-DOPA. It is likely that cutaneous catecholaminergic system will communicate with brain by activating sensory nerves, or, with other tissues, via entry into systemic circulation and by affecting immune cells circulating from the skin to other organs (Fig. 1.1).

2.3 Histamine

2.3.1 Production and Metabolism of Histamine

Histamine is derived from the decarboxylation of histidine by the L-histidine decarboxylase. After release, histamine is degraded by histamine-*N*-methyltransferase (in brain and at periphery) or diamine oxidase (in the periphery) (Fitzpatrick et al. 1993; Zhang et al. 2007). Histamine is produced mainly by mast cells and basophils. Cross-linking of IgE antibodies attached to the cell membrane represents a main mechanism for histamine release. Histamine binds to four different types of seven-transmembrane receptors that signal through G-proteins.

The H_1 receptor is found on smooth muscle and endothelial cells and is responsible for smooth muscle contraction and decreased adhesion of endothelial cells. H_2 receptor is located on vascular smooth muscles and parietal cells in the stomach and is responsible for vasodilatation and gastric acid secretion. H_3 receptor is found in the central and peripheral nervous systems and is responsible for decreased secretion of several neurotransmitters including histamine, acetylocholine, serotonin, and norepinephrine. H_4 receptor is found primarily on basophils and has a role in chemotaxis (Fitzpatrick et al. 1993; Zhang et al. 2007).

2.3.2 Bioregulatory Role of Histamine in the Skin

In the epidermis, H_1 and H_2 receptors are expressed on keratinocytes (Albanesi et al. 1998; Koizumi and Ohkawara 1999; Koizumi et al. 1998; Shinoda et al. 1998) and H_2 receptors on epidermal melanocytes (Yoshida et al. 2000). Mediators released from mast cells inhibit keratinocyte growth in culture (Huttunen et al. 2001). Activation of keratinocyte H_2 receptors affects proliferation and differentiation via activation of the cyclic AMP pathway and also phospholipase C pathway with associated increase in intracellular calcium levels (Koizumi and Ohkawara 1999). In mouse keratinocytes, H_2 receptor signaling through the PLC second messenger system is inhibited during calcium-induced keratinocyte differentiation by an autocrine loop which involves downregulation of H_2 receptor expression and inhibition of histamine metabolism (Fitzsimons et al. 2002). In keratinocytes, activation of the H_1 receptor enhances UVB-induced IL-6 production (Koizumi and Ohkawara 1999; Koizumi et al. 1998), whereas H_1 receptor antagonists inhibit ICAM-1 expression (Ling et al. 2004). Histamine upregulates keratinocyte MMP-9 production via the H_1 receptor (Gschwandtner et al. 2008). H_2, however, not H_1, agonists stimulate intracellular calcium signaling in keratinocytes (Koizumi and Ohkawara 1999). In these cells, histamine acting on H1 receptors increases the expression of IFN-γ-induced intercellular adhesion molecule 1 (ICAM-1) and MHC class I molecules. It also augments IFN-γ-induced release of chemokines such as CXCL10, as well as the release of GM-CSF via protein kinase Cα and extracellular signal-regulated (ERK) kinase (Giustizieri et al. 2004; Kanda and Watanabe 2004). In cultured keratinocytes, histamine through the activation of H_1 receptor inhibits CCL17 production by suppressing p38 MAP kinase and NF-κB activities. Histamine acts as a negative feedback signal for existing Th2-dominant inflammation by suppressing CCL17 and enhancing CXCL10 production (Fujimoto et al. 2011). The effect of histamine acting through H2 receptor appears to be the opposite. Histamine, via H_2 receptor, increases survival of keratinocytes acting by NF-κB activation (Kim and Lee 2010). IL-17, produced by Th17 cells infiltrating into the dermis (a cytokine involved in various inflammatory skin diseases including psoriasis), stimulates keratinocytes to produce inflammatory mediators such as IL-36, TNF-α, IL-6, and IL-8 (Carrier et al. 2011). Histamine markedly augments the production of IL-8 and GM-CSF in the presence of IL-17 and TNF-α in keratinocytes (Moniaga et al. 2011). Moreover, histamine induces human β-defensin 2 and 3 production in keratinocytes acting via H_1 receptors by activating NF-κB, AP-1 pathway, or STAT1, STAT3, and AP-1 as well as JAK2 and MEK/ERK signaling pathways (Ishikawa et al. 2009; Kanda and Watanabe 2007). Histamine promotes cutaneous antimicrobial defenses and wound repair by stimulating secretion of defensins (Ishikawa et al. 2009; Kanda and Watanabe 2007). Histamine also enhances nerve growth factor production by inducing c-Fos expression in keratinocytes (Kanda and Watanabe 2003).

The activation of the H_2 receptors on melanocytes stimulates melanogenesis (Yoshida et al. 2000). Histamine, similarly to α-MSH, contributes to hyperpigmentation by enhancing eumelanin/pheomelanin ratio (Lassalle et al. 2003). Acting at the H_2 receptor histamine stimulates melanocyte migration in culture via signaling through ERK, CREB, and Akt (Kim and Lee 2010). Histaminergic system is upregulated in the B16F10 melanoma cells when compared to noncancerous melanocytes, which indicates that it might have a role in tumorigenesis (Davis et al. 2011). Both Western blot and immunohistochemical studies showed much stronger histidine decarboxylase expression in melanoma cells as compared to normal melanocytes (Haak-Frendscho et al. 2000). Moreover, H_1 histamine receptor antagonists were shown to induce genotoxic and caspase-2-dependent apoptosis in human melanoma cells, but not normal melanocytes (Jangi et al. 2006).

In the dermis, histamine receptors are expressed on fibroblasts, immunocytes, endothelial cells, blood vessels, smooth muscle, and nerve endings (Fitzpatrick et al. 1993). In Th2 lymphocytes stimulation of H_4 receptor led to the activation of transcription factor AP-1 followed by the release of IL-31, which is involved in the development of pruritus (Gutzmer et al. 2009). On the other hand, activation of H_4 histamine receptors expressed on monocytes activated intracellular calcium mobilization and inhibited the CCL2 chemokine production which reduced recruitment of monocytes (Dijkstra et al. 2007). Histamine acts on H_4 receptors of eosinophils and mediates their chemotaxis, induces cell shape change, and upregulates adhesion molecules CD11b/CD18 (Mac-1) and CD54 (ICAM-1). This effect, while observed in cultured eosinophils, may be of paramount importance in the skin (Ling et al. 2004).

Histamine also acts on H_2 and H_4 receptors of plasmacytoid dendritic cells and downregulates production of TNF-α, IFN-α, and CXCL8 (Mazzoni et al. 2003). Plasmacytoid dendritic cells migrate in response to H_4 receptor agonist stimulation. Of note, H_4 receptor is present in high levels on plasmacytoid dendritic cells in the lesional psoriatic skin (Gschwandtner et al. 2011).

2.3.3 Conclusions

Histamine is produced not only by mast cells but also by other cells of epidermis and dermis and acts locally in the epidermis and dermis by binding to H_1-H_4 receptors. Histamine targets not only endothelium and smooth muscles of blood vessels but also modulates function of keratinocytes, melanocytes, and cells of skin immune system. It affects intracellular signaling cascades, cell proliferation, and melanogenesis. Histamine is upregulated in melanoma cells. It signals mainly via H_4 receptor on the cells of the immune system and affects their migration and cytokine secretion patterns. Moreover, it modulates Th2-type immune responses and antimicrobial peptide expression. Thus, histamine is an important part of the neuro-immuno-endocrine system of the skin (Slominski and Wortsman 2000) with local and systemic effects (Figs. 1.1 and 1.2).

2.4 Serotoninergic System

2.4.1 Production and Metabolism of Serotonin

2.4.1.1 An Overview

Serotonin (5-hydroxytryptamine, 5-TH) is widely synthesized throughout the animal kingdom, plants, and unicellular organisms (Azmitia 2001, 2007). In plants, serotonin serves as a trophic factor and an antioxidant which is similar to the animal kingdom (Azmitia 2001). In humans, serotonin was shown to be synthesized predominantly by intestinal enterochromafin cells with other sites of production represented by the central nervous system, pineal gland, retina, ovaries, placenta, thymus, pancreas, skin, breast, gestational tissues, blood vessels, rectal epithelium, bronchial epithelial cells, thyroid parafollicular cells, mast cells, and lymphocytes (Nordlind et al. 2008).

The first obligatory step in the synthesis of serotonin is the hydroxylation of L-tryptophan to produce 5-hydroxytryptophan (TrpOH) in a reaction catalyzed by tryptophan hydroxylase (TPH) (Mockus and Vrana 1998), a protein encoded by either TPH1 gene expressed ubiquitously (Mockus and Vrana 1998) or TPH2 gene expressed predominantly in the brain (Zhang et al. 2004). This reaction requires oxygen and cofactor 6BH4. TrpOH is further decarboxylated by AAD to produce 5-HT. In humans, L-tryptophan is present in blood plasma at steady-state level both in the free form (approximately 1.2×10^{-5} M) and bound to serum albumins (ca. 6×10^{-5} M), with TPH having a Kda for tryptophan of approximately 10^{-8} M. Thus, fluctuations in free pool of tryptophan directly and immediately alter the level of serotonin synthesis (Nordlind et al. 2008). Catabolism of serotonin is initiated by MAO with the production of 5-hydroxyindoleacetaldehyde, oxidized further by aldehyde dehydrogenase (E.C. 1.2.1.3) to 5-hydroxyindole-3-acetic acid (5-HIAA), which is the main product of metabolism, or reduced to 5-hydroxytryptophol (HTOL) by alcohol dehydrogenase (E.C. 1.1.1.1) (Fig. 2.3). 5-HT can also be methylated to 5-methoxytryptamine (5MTT) and catabolized as shown in Fig. 2.3. Additional pathway involves serotonin acetylation by arylalkylamine N-acetyltransferase (AANAT) or arylamine N-acetyltransferase isoenzyme showing substrate specificity toward both arylamines and arylalkylamines to produce N-acetylserotonin (NAS)

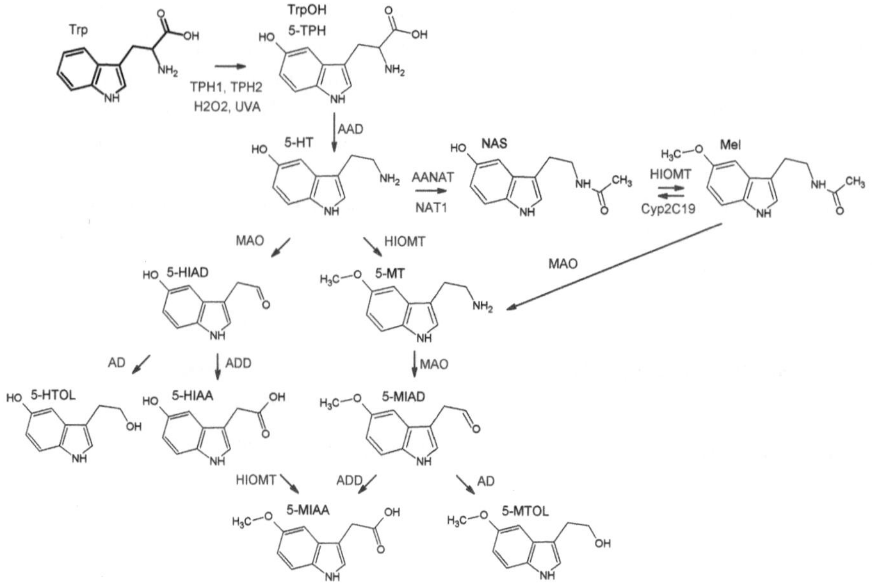

Fig. 2.3 Biochemical pathway of serotonin synthesis and metabolism in the skin. The pathway starts with hydroxylation of tryptophan by tryptophan hydroxylase type 1 or 2 (TPH1 or TPH2) to form 5-hydroxytryptophan (5-TPH; TrpOH). TrpOH can also be produced by nonenzymatic action of UVA and H_2O_2. Serotonin (5-hydroxytryptamine, 5-HT) derives from 5-TPH by action of L-amino acid decarboxylase, AAD. Serotonin can be acetylated by aralkylamine N-acetyltransferase (AANAT) or N-acetyltransferase (NAT) to produce N-acetylserotonin (NAS) with further methylation by hydroxy-indole-O-methyl transferase (HIOMT) to melatonin. Deactivation of serotonin is catalyzed mainly by MAO with the formation of 5-hydroxyindoleacetaldehyde (5-HIAD) which is followed by the action of alcohol (AD) or aldehyde dehydrogenase (ADD) with the formation of 5-hydroxytryptophanol (5-HTOL) or 5-hydroxyindole-3-acetic acid (5-HIAA), respectively. Alternatively, HIOMT activity may also lead to the production of methylated derivatives of serotonin. The first step catalyzed by HIOMT leads to the formation of 5-methoxytryptamine 5-MT. The subsequent action of MAO results in 5-metoxyindoleacetaldehyde (5-MIAD) formation. Finally, AD or ADD facilitates the synthesis of 5-methoxytryptophol (5-MTOL) or 5-methoxyindole-3-acetic acid (5-MIAA), respectively. HIOMT was found also to catalyze the conversion of 5-HIAA to 5-MIAA. By the action of MAO melatonin can be metabolized to 5-methoxytryptamine (5-MT), thus entering the pathway leading to 5 MTOL or 5 MIAA formation

(Fitzsimons et al. 2002; Klein 2004). NAS can also be further metabolized to melatonin (Reiter 1991). In the skin, a number of NAS metabolites unrelated to melatonin were found, the nature and mechanism of generation of which remain to be defined (Slominski et al. 2003b, c). After release into blood, serotonin is actively taken up into platelets and stored in solid granules with the help of a serotonin transporter (5-HTT), a member of the Na^+/Cl^--dependent transporter superfamily, which actively regulates serotonin transport. Serotonin can be transported through the plasma membrane in either direction; however, under most conditions, its reuptake is favored (Nordlind et al. 2008). Plasma serotonin is also cleared by the liver and lung endothelial cells and further catabolized to 5-HIAA.

2.4.1.2 Production and Metabolism of Serotonin in the Skin

Mammalian skin cells can produce serotonin via the sequential transformation of L-tryptophan by TPH and AAD (Slominski et al. 2005c) (Fig. 2.3). Thus, the *TPH1* gene is expressed in human skin under normal and pathological conditions as well as in a wide array of normal and transformed human epidermal, dermal, and adnexal skin cells with some cells expressing the aberrant *TPH1* transcript (Slominski et al. 2002c, 2003b, c). As to the *TPH2* gene, it is expressed in the retinal pigment epithelium (Zmijewski et al. 2009b) and normal and malignant melanocytes (Zmijewski and Slominski, unpublished). Although the *TPH* gene is expressed almost in all types of human skin cells, the highest expression was found in normal and malignant melanocytes that also accumulated significant amounts of serotonin (Figs. 2.4 and 2.5) (Slominski et al. 2003a, 2005c). Interestingly, the enzymatic conversion of tryptophan to TrpOH in melanoma cells occurs at high levels, comparable to those in the brain (Slominski et al. 2002a, c). TPH and *TPH1* were also detected in the mouse and hamster skin, and in cultured mouse follicular melanocytes and melanoma cells (Slominski et al. 2002a, 2003b, c). Interestingly, the *TPH1* gene expression changes during murine hair cycle (Slominski et al. 2003b, c). In addition, TPH and serotonin are strongly expressed in rodent masts cells. It is also important to notice that the skin has a capability for de novo synthesis/recycling of the 6BH4 (Schallreuter et al. 1997, 1998, 2008a) and of pyridoxal $5'$-phosphate (PLP) (Coburn et al. 2003) both serving as important cofactors necessary for the production of TrpOH and serotonin. Interestingly, nonenzymatic production of TrpOH through H_2O_2 and UVA radiation indicates that a free-radical-mediated oxidation of L-tryptophan is also possible in the skin (Schallreuter et al. 2008a).

In human skin biopsies immunoreactivity of TPH and serotonin was found in normal epidermal melanocytes and malignant melanomas (Figs. 2.4 and 2.5) (Slominski et al. 2003a) with additional detection by immunofluorescence techniques in epidermal keratinocytes, hair follicles, eccrine glands, blood vessels, and skin mast cells (Slominski et al. 2005c). These findings are consistent with the immunodetection of serotonin in perivascular human mast cells of adrenal cortex (Lefebvre et al. 2001) and breast epithelial cells (Matsuda et al. 2004). Serotonin was also detected by immunocytochemistry in dermal Merkel cells in rat and pig skin at the epidermal rete ridges and upper hair follicles adjacent to nerve terminals (Nordlind et al. 2008). Cutaneous serotonin content can be affected by inflammatory processes (Lonne-Rahm et al. 2008; Nordlind et al. 2008; Rasul et al. 2011; Thorslund et al. 2009). For example, human skin affected by psoriasis or chronic eczema showed elevated expression of serotonin in the epidermal and adnexal structures (Nordlind et al. 2008).

The catabolism of serotonin in mouse skin is initiated by its deamination by MAO, followed by the oxidation or reduction of the resultant 5-hydroxyindole acetaldehyde to 5-HIAA and/or 5-HTPOL (Slominski et al. 2003b, c). Similar metabolism was uncovered in rat skin, although in this species 5-HIAA was the

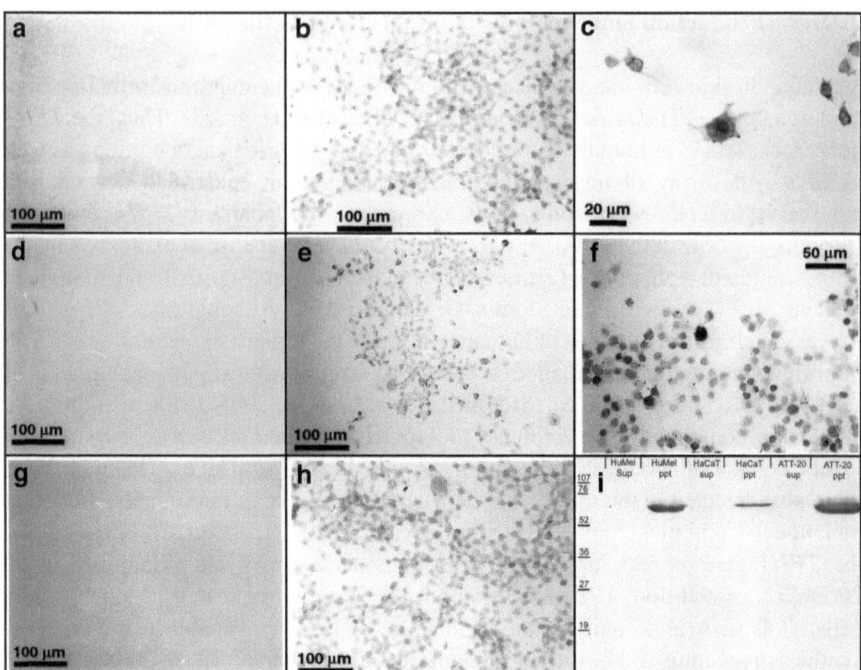

Fig. 2.4 Expression of TPH, serotonin (5-HT), and serotonin transporter (5-HTT) in skin cells. Panels **a–h** show immunocytochemical detection of 5-HT (**b, c**), 5-HTT (**e, f**), and TPH (**h**) in fixed cells using corresponding antibody at the dilution of 1:5,000 (antibody against 5-HT, Diasporin Corp., Stillwater, MN) or 1:1,000 (antibodies against TPH and 5-HTT, Chemicon, Temecula, CA). (**a, d, g**) Negative controls incubated with secondary antibody only. (**I**) Western blot showing detection of 5-HTT in membranous (ppt) but not cytosolic (sup) fractions from human melanoma (HuMel), HaCaT keratinocytes (HaCaT), and ATT-20 pituitary cells. For technical details of the assays, see Slominski et al. (2005d)

main degradation product and 5-HTPOL remained below the limit of detectability (Semak et al. 2004). MAO metabolism of serotonin was also detected in guinea pig skin (Tachibana et al. 1990) and production of 5-HIAA was documented in human epidermal keratinocytes and melanoma cells (Slominski et al. 2002c).

The alternative serotonin metabolism pathway in the skin is represented by its acetylation to *N*-acetylserotonin, which in human and rodent skin and cultured skin cells is mediated via the action of either AANAT or NAT with mixed arylamine/ arylalkylamine substrate specificity (Slominski et al. 2005c). In hamster skin, we characterized two *N*-acetyltransferase activities including NAT-1 with substrate specificity toward arylamines, and NAT-2 showing substrate specificity toward both arylamines and arylalkylamines such as serotonin, tryptamine, and methoxytryptamine (Gaudet et al. 1993; Slominski et al. 2002a). Furthermore, we demonstrated that at least part of this activity in hamster, rat, and human skin represented native AANAT (Slominski et al. 2002a). In accordance, serotonin *N*-acetyltransferase activity was significantly inhibited by low concentrations of

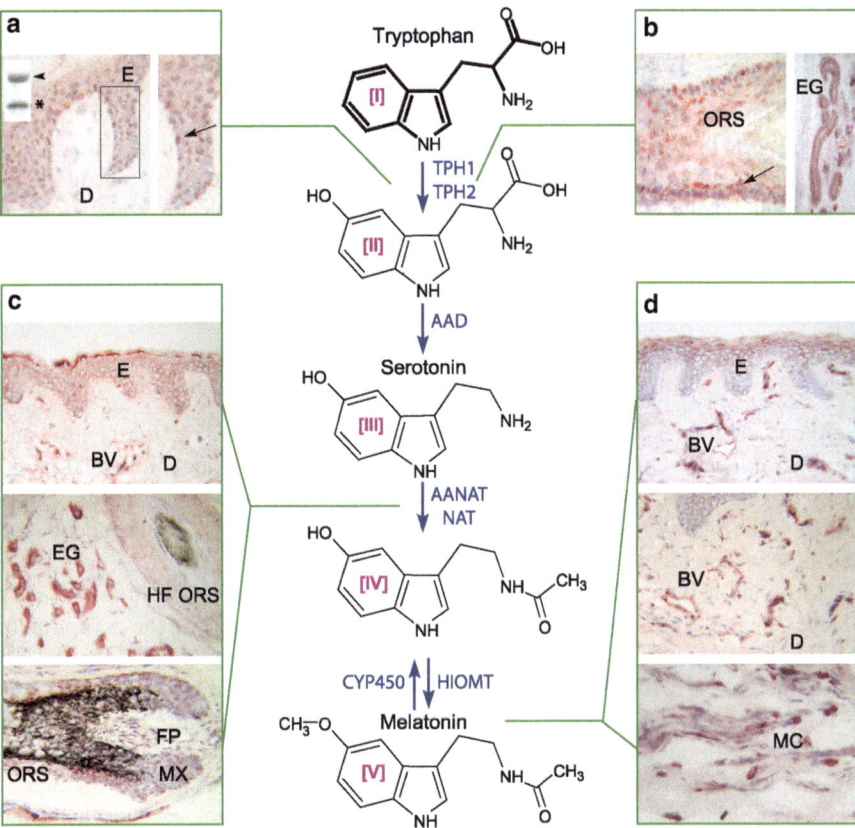

Fig. 2.5 Melatoninegic system in the skin. TPH1 Western blot insert in the Panel **a** is of approximately 50 kDa (*arrowhead*) that is processed and/or degraded to lower molecular weight species (*asterisk*). It was immunolocalized in the epidermis (ES), hair follicle (ORS), eccrine glands (EG), showing the highest expression in melanocytes (*arrows*) (Panels **a** and **b**). 5-hydroxytryptophan is further decarboxylated by aromatic amino acid decarboxylase (AAD). AANAT (enzyme acetylating serotonin) is expressed in cells of epidermal, dermal, and adnexal compartments (E, BV, EG, and hair follicle structures in Panel **c** on the *left*). Immunocytochemical localization of melatonin-like immunoreactivity is shown in Panel **d** on the *right* (upper E, BV, and MC). Immunocytochemistry was performed on human skin biopsies: *E* epidermis, *D* dermis, *BV* blood vessel, *EG* eccrine gland, *HF ORS* hair follicle outer root sheath, *FP* hair follicle papilla; *MX* hair follicle matrix, *MC* mast cells. For technical details, see Slominski et al. (2005d). Reproduced with permission from the publisher (Slominski et al. 2008a)

coenzyme A-S-N-acetyltryptamine [Cole bisubstrate; BSI, see (Hickman et al. 1999; Khalil et al. 1998)], indicating true AANAT activity. However, significant enzymatic activity generating NAS was resistant to BSI suppression, showing that in rodents arylamine activity (NAT-2) resistant to BSI can also participate in the acetylation of serotonin (Semak et al. 2004; Slominski et al. 2002a). Rodent NAT-2 is a homologue of human NAT-1; thus, it is likely that NAT-1 may contribute to NAS production also in the human skin. Interestingly, in the C57BL/6 mouse producing inactive AANAT (Roseboom et al. 1998), we detected cutaneous trans-formation of serotonin to NAS and, to a lesser extent, acetylation of tryptamine (Slominski et al. 2003b, c). Most interestingly, acetylation of serotonin, but not of tryptamine, was dependent on the phase of hair cycle, skin anatomic location, and the presence of pathology (melanoma). NAS was further metabolized to several products (the chemical nature of which remains to be defined) in a hair cycle-dependent fashion (Slominski et al. 2003b, c). In humans, both skin racial pigmen-tation and cutaneous pathology determine the reaction rate and specificity of serotonin acetylation (Slominski et al. 2002c).

2.4.2 Bioregulatory Role of Serotonin in the Skin

Serotonin regulates a wide range of physiological processes at the central and peripheral levels acting as a neurotransmitter, hormone, cytokine, biological modi-fier, growth factor, morphogen, and antioxidant or pro-oxidant (Azmitia 2007, 2010). The above functions are mediated through receptor-dependent and receptor-independent mechanisms (Hoyer et al. 2002).

Serotonin acts via multiple receptor subtypes labeled as 5-HT1 through 5-HT7 (Hoyer et al. 2002). Most of these receptors are metabotropic, with the exception of 5-HT3, which is ionotropic and primarily gates sodium and potassium ions. 5-HT1 receptors (1A, 1B, 1D, 1E, and 1F) couple via Giα to inhibit cAMP formation while 5-HT4, 5-HT6, and 5-HT7 all couple via Gsα to stimulate cAMP production (Hoyer et al. 2002). In addition, 5-HT1A receptors produce membrane hyperpolarization by coupling to K^+-channels. 5-HT2 (2A, 2B, and 2C) receptors couple via Gqα to phosphatidylinositol hydrolysis and the formation of inositol trisphosphate and diacylglycerol (Hoyer et al. 2002). The 5-HT5 receptor (5A and 5B) is considered to be an orphan receptor. Serotonin receptor function can be modulated by RNA editing, endogenous lipids that act as allosteric modulators, and serotonin moduline (tetrapeptide, 5-HT-moduline) that is produced by proteolytic modification of chromogranin. 5-TH moduline is an allosteric modulator which regulates 5-HT5 receptor dimerization and formation of either homodimers or heterodimers. The receptors' heterogeneity and functional diversity are also amplified by the process of alternative splicing and differential subunit incorporation into the receptor complex. The regulation of 5-HT receptor activity is also affected by serotonin transporters, which remove serotonin from the extracellular environment or, under certain conditions, pump it out of the cell.

In human skin and skin cells, we identified expression of genes coding 5-HT receptors, including *HTR1A, 1B, 2A, 2B, 2C*, and *7* genes, and it was shown that the pattern of expression was cell type specific and modified by skin pathology (Slominski et al. 2003d). Interestingly, alternatively spliced form of *HTR2C* with a deletion of exon 2, a fragment of exon 3, and an insertion of cryptic exon containing termination codon was found in human melanoma, while the *HTR2B* isoform with a deletion of exon 2, but with a preserved reading frame coding for a receptor protein without transmembrane domains 3 and 4 was found in normal human skin and skin affected by basal cell carcinoma (Slominski et al. 2003a). We also found RNA editing (A to G substitution) in human *HTR7* gene (Slominski et al. 2003a), which may be connected to the local expression of adenosine deaminases. In mouse and hamster skin, expression of the *HTR2B* and *HTR7* genes was demonstrated, which was dependent on the phase of hair cycle (mouse) and type of tissue or cells (Slominski et al. 2004b).

We should also mention that Kaneko et al. have failed to detect 5-HT2A gene in epidermal keratinocytes (Kaneko et al. 2009). However, these findings have to be considered with caution, since other researchers demonstrated that 5-HT2A antagonists inhibited UVR-induced skin carcinogenesis (Sreevidya et al. 2008, 2010) and that sunlight-induced immunosuppression could be mediated via the activation of 5-HT2A by cis-urocanic acid (Walterscheid et al. 2006). Furthermore, 5-HT2A protein was detected by immunocytochemistry in dermal lymphocytes, fibrocytes, vasculature, and sensory nerve endings, abating the epidermis (Nordlind et al. 2008), while 5-HT1A receptor was localized to keratinocytes of the upper epidermis, epidermal melanocytes, mast cells, and dermal vasculature (Nordlind et al. 2008). Furthermore, 5-HT1A and 5-HT2A were detected in the majority of benign tumors such as compound nevi, dysplastic nevi, and also in malignant melanomas (Nordlind et al. 2008). By the use of immunocytochemistry, 5-HT2C was detected in epidermal Langerhans cells and melanocytes, 5-HT3 in the basal epidermal keratinocytes, and 5-HT7 in dermal vasculature (Nordlind et al. 2008). 5-HT1A, 2A, and 2C were also detected in rodent skin dermal and epidermal immune cells (Nordlind et al. 2008). Diverse expression of 5-HT receptors was also found in immune cells that were dependent on cell type and their level of activation.

Also Merkel, Langerhans and mast cells, lymphocytes and macrophages (Nordlind et al. 2008), and immortalized human epidermal keratinocytes and melanoma cells express 5-HTT (Fig. 2.4). Their role is substantiated by observations which showed that serotonin uptake inhibitors could induce spontaneous bruising, pruritus, urticaria, angioedema, erythema multiforme, the Steven–Johnson syndrome, toxic epidermal necrolysis, erythema nodosum, alopecia, hypertrichosis, leukocytoclastic vasculitis, and acneiform eruption (reviewed by Nordlind et al. 2008; Slominski et al. 2005c). This can also be associated with flares of psoriasis vulgaris and development of delayed hypersensitivity.

Under in vitro conditions, serotonin exerted variable effects on skin cells depending on the context (Nordlind et al. 2008; Salim and Ali 2011; Slominski et al. 2005c). It stimulated proliferation of dermal fibroblasts (Slominski et al. 2005c), similarly to non-skin fibroblasts (Seuwen and Pouysségur 1990). Serotonin also stimulated growth of epidermal melanocytes in the absence of growth factors, while inhibiting their proliferation in media supplemented with serum (Slominski et al. 2003a). The former effect could be linked with the stimulation of intracellular cAMP accumulation, while the latter could represent serotonin antagonism with serum growth factors (Slominski et al. 2005c). NAS, the product of serotonin metabolism, showed no effect on the proliferation of fibroblasts and melanocytes (Slominski et al. 2003a) and serotonin or inhibitors of its uptake inhibited melanogenesis (reviewed by Slominski et al. 2004b; Slominski et al. 2005c). In addition, serotonin modulated proliferation of cultured murine keratinocytes (Maurer et al. 1997). Interestingly, serotonin content within mast cell granules steadily decreased throughout anagen and increased during catagen and telogen phases of hair cycle (Hasse et al. 2007).

Serotonin shows vasoactive and immunomodulatory effects. For example, it plays a role in the Arthus reaction (Tachibana et al. 1990; Yuasa et al. 2001), induces sustained vascular permeability (Fujii et al. 1994), and also modulates the inflammatory response to substance P (SP) via capsaicin-sensitive sensory fibers (Khalil and Helme 1990). Serotonin participates in the activation of T cells and natural killer cells by macrophages, initiation of delayed-type hypersensitivity responses, production of chemotactic factors, and the modification of innate immune responses (Benton et al. 2010; Betten et al. 2001; Cloez-Tayarani and Changeux 2007; Hsueh et al. 2002; Mossner and Lesch 1998). In allergic contact dermatitis and psoriasis, the number of cells expressing both 5-HT1A and tryptase diminishes, whereas the number of dermal cells expressing 5-HT2A and CD3 increases, including atopic dermatitis (Lonne-Rahm et al. 2008; Nordlind et al. 2008; Rasul et al. 2011; Thorslund et al. 2009). Similar pattern is found in the murine epidermis affected by contact eczema. Furthermore, both eczematous and psoriatic human skin shows increased number of mononuclear cells expressing 5-HTT (reviewed by Nordlind et al. 2008). In addition, serotonin can act as a chemoattractant for eosinophils, probably by binding to 5-HT2A receptors. It is involved in the mast cell recruitment to the site of tissue injury through the activation of 5-HT1A, however, without inducing their degranulation (Nordlind et al. 2008). Regulatory function of 5-HT1A in inflammatory responses is emphasized by the suppression of the severity of contact allergy in rats, after topical or oral administration of its agonist, buspirone (Nordlind et al. 2008). Another 5-HT1A agonist, tandospirone, attenuates itching in patients with atopic dermatitis (Nordlind et al. 2008). On the other hand, treatment with 5-HT2A antagonists reduced the severity of contact allergic reactions in mice and one of them, spiperone, was effective when applied either systemically or topically. Furthermore, 5-HT2 receptor antagonist, ketanserin, inhibited the established but not challenge-induced phases of allergic contact dermatitis (Nordlind et al. 2008). Serotonin is also involved in the pathogenesis of cholestatic and uremic pruritus, urticaria, and itch reaction (reviewed by Slominski et al. 2005c).

2.4.3 Serotonin Receptors on Sensory Nerves

5-HT receptors were widely detected on cutaneous sensory nerve endings (reviewed by Nordlind et al. 2008; Slominski et al. 2005c). Intradermal injection of serotonin into rat elicited enhanced c-fos-like immunoreactivity in superficial lamina at the lateral aspect of the dorsal horn, in a manner similar to the immunoreactivity evoked by capsaicin. The 5-HT receptor was detected in unmyelinated sensory axons at the dermal–epidermal junction and the nerve endings of Pacinian corpuscles of rat glabrous skin (Carlton and Coggeshall 1997) and rat sinus hair follicle (Tachibana et al. 2005). 5-HT1 receptors are present in the dermis of rabbits on afferent nerve fibers around hair follicles and sebaceous glands (Branchek et al. 1988). 5-HT2A receptors are partially responsible for mediating scratching in mice (Tachibana et al. 1990). Although neither 5-HT2 nor 5-HT3 appears to be involved in itch responses caused by chronic allergic skin dermatitis in rats, acute scratching is mediated by skin 5-HT2 receptors, and intradermal injection of serotonin induced itching in normal, but not inflamed skin (reviewed by Nordlind et al. 2008; Slominski et al. 2005c). In human skin, 5-HT2A and 5-HT3 are localized on sensory nerve ending in the dermis or located close to or entering the epidermis, and their activation may explain pruritic responses to intradermally injected serotonin (Nordlind et al. 2008; Slominski et al. 2005c). Specifically, an antagonist of 5-HT3, ondansetron, can reduce the severity of pruritus, while paroxetine is used in the treatment of pruritus and its antipruritic action is connected with downregulation of 5-HT3 expression (Nordlind et al. 2008; Slominski et al. 2005c).

2.4.4 Reception of Ultraviolet Light

The cutaneous serotoninergic system may play a role in body reception of and reaction to light (Slominski et al. 2005c). For example, it has been reported that UVA-induced well-being can be linked to increased serum serotonin and decreased melatonin levels after a single radiation exposure (Gambichler et al. 2002). It has also been proposed that 5-HT2A plays a role in the transduction of UVR energy into biological responses by serving as the receptor for cis-urocanic acid (cis-UCA), generated through photoisomerization of the trans-UCA in the stratum corneum after absorption of UVR (Walterscheid et al. 2006). Cis-UCA acts as a powerful local and systemic immunosuppressor (Garssen et al. 2001), and it was proposed that 5-HT2A mediates immunosuppressive effects of UVR after binding of cis-UCA (Walterscheid et al. 2006). A role for 5-HT2A in UVB-induced skin photocarcinogenesis was also suggested (Sreevidya et al. 2008, 2010). Other authors proposed that cis-UCA and serotonin mediate UVB-induced immunomodulation, however, via independent pathways in which cis-UCA does not act through 5-HT2A (Kaneko et al. 2009). Thus, there is sufficient information to support involvement of the local serotoninergic system in cutaneous responses to the UV light; however, the mechanism may be more complex than originally anticipated. It may include activation of 5-HT receptor signaling on either nerve ending or skin cells secondary to UVR-induced local production of serotonin or alternative ligands for HT receptors with a consequent regulation of local homeostasis and immune system. Such signals will be projected to the brain via the ascending nerve routes. Furthermore, release of serotonin into circulation may generate endocrine effects.

2.4.5 Conclusions

The mammalian skin cells have the capability to produce and metabolize serotonin. The cutaneous phenotypic effects are mediated by its interactions with 5-HT receptors including 5-HT1A, 1B, 2A, 2B, 2C, 3 and 7, and 5-HTT receptors, which are expressed in a cell type-dependent manner. The serotonin receptors are also expressed on sensory nerve endings, which transmit to the brain information on changes in skin homeostasis induced by either intrinsic or environmental factors (Slominski 2005; Slominski and Wortsman 2000). The topical application of specific receptors agonists or antagonists, serotonin uptake inhibitors or modulation of local serotonin production/degradation may represent future novel therapies of skin diseases including neurodermatoses and itching disorders. Finally, the cutaneous serotoninergic system may be involved in the transformation of light energy of solar radiation into local and systemic biological responses, with the latter mediated via transmission to brain, endocrine effects, or regulation of systemic responses as shown on Figs. 1.1 and 1.2.

Chapter 3
Melatoninergic System in the Skin

3.1 Melatonin Production

Melatonin production is highly conserved in nature through different species including bacteria, unicellular eukaryotes, algae, plants invertebrates, and vertebrates (Hardeland et al. 2011; Reiter 1991; Slominski et al. 2008a; Tan et al. 2002; Yu and Reiter 1993). In mammals, melatonin is produced in the pineal gland (Reiter 1991) as well as in brain, retina, Harderian gland, ciliary body, lens, thymus, airway epithelium, bone marrow, immune cells, gonads, placenta, gastrointestinal tract, and skin (Bubenik 2002; Carrillo-Vico et al. 2004; Hardeland et al. 2011; Kanda and Watanabe 2007; Pandi-Perumal et al. 2006; Slominski et al. 2005a, 2008a; Watson 1994; Zmijewski et al. 2009b), and perhaps other organs. Circulating melatonin predominantly derives from the pineal gland by diffusion into the circulation, although entry from other extra-pineal sites of production is also possible.

Melatonin is a product of a two-step transformation of serotonin which involves acetylation catalyzed by AANAT to NAS (a rate-limiting step) followed by methylation by hydroxyindole-O-methyltransferase (HIOMT, EC 2.1.1.4) to produce melatonin (N-acetyl-5-methoxytryptamine) (Reiter 1991; Yu and Reiter 1993). In the pineal gland melatonin production is controlled by the suprachiasmatic nucleus through nocturnal sympathetic release of norepinephrine that acting via adrenergic receptors activates cAMP-dependent signal transduction cascades leading to the stimulation of AANAT and ultimate production of melatonin (Klein 2007; Reiter 1991; Yu and Reiter 1993). Melatonin synthesis is also potentiated by vasoactive intestinal peptide (VIP), pituitary adenylate cyclase-activating peptide (PACAP), and neuropeptide Y (Klein 2007; Reiter 1991; Yu and Reiter 1993). NAS can also be produced by the action of arylamine N-acetyltransferases (Fitzsimons et al. 2002) as it was shown in human (Slominski et al. 2002c), rat (Semak et al. 2004), hamster (Gaudet et al. 1993; Slominski et al. 2002a), and murine skin (Slominski et al. 2003b, c). NAS can be further methylated to melatonin, depending on the anatomic location and activity of HIOMT. This is best illustrated in the C57BL/6

A.T. Slominski et al., *Sensing the Environment: Regulation of Local and Global Homeostasts by the Skin's Neuroendocrine System*, Advances in Anatomy, Embryology and Cell Biology 212, DOI 10.1007/978-3-642-19683-6_3, © Springer-Verlag Berlin Heidelberg 2012 mice, defined by some authors as a natural melatonin "knockdown" (Kobayashi et al. 2005; Roseboom et al. 1998; Slominski et al. 2003b, c). Specifically, in the C57BL/6 mouse serotonin can be acetylated to NAS in a reaction mediated by an enzyme different from conventional AANAT providing an important mechanistic explanation for the significant production of melatonin in the peripheral organs of this species, which express HIOMT (Ma et al. 2008; Scarparo et al. 2000; Slominski et al. 2003b, c). In addition, the existence of low flux rate alternative pathways has been proposed that involves *O*-methylation of serotonin with subsequent *N*-acetylation, or *O*-methylation of tryptophan followed by consecutive decarboxylation and *N*-acetylation.

Transcripts of *AANAT* and of *HIOMT* genes were detected in normal and pathological skin biopsies, and in most skin cells cultured in vitro including normal keratinocytes (neonatal and adult, epidermal and follicular), immortalized HaCaT keratinocytes, fibroblasts (dermal and hair follicle papilla), normal melanocytes, several melanoma cell lines, and squamous cell carcinoma cells (Slominski et al. 2002b). Interestingly, novel isoforms of *AANAT* and *HIOMT* were detected in normal and pathological skin (invaded by basal cell carcinoma cells) and in neonatal keratinocytes (Slominski et al. 2002b). Gene expression in epidermal and dermal skin cells was followed by the synthesis of the AANAT and HIOMT enzymes with the detection of corresponding enzymatic activities (Slominski et al. 2002b).

The acetylation of serotonin was also dependent on local cellular environment. Thus, when AANAT activity was calculated for two substrates, tryptamine and serotonin, the activity ratios were close to 1 for all melanoma lines and for keratinocytes. On the other hand, these ratios ranged from 2.5 to 6 for whole skin from three white subjects and zero in melanocytes and in whole skin of a black subject whose AANAT activity toward tryptamine was below detectability level. These findings suggest a role for both skin racial pigmentation and type of cutaneous pathology (such as melanoma) in this regulation (Slominski et al. 2002b). Both of them may be important determinants of reaction rate and specificity of serotonin acetylation. Using immunocytochemistry AANAT antigen was detected in suprabasal differentiating keratinocytes in human scalp epidermis. However, melanocytes also exhibited immunoreactivity for this enzyme (Fig. 2.5). High expression of the antigen was also seen in the outer peripheral epithelial layers of the anagen hair follicles (Fig. 2.5) and the basal cells of the sebaceous and eccrine glands. The expression was further found in sensory nerve endings abutting the epidermal layers (Slominski et al. 2005c). Melatonin-like immunoreactivity in human skin was detected on differentiating keratinocytes in spinous and granular layers of the epidermis (Fig. 2.5). The antigen was not expressed in keratinocytes of basal and suprabasal layers of the epidermis, while being found in singly scattered melanocytes. Melatonin immunoreactivity was also detected throughout the hair follicle epithelium, in blood vessels, and cutaneous mast cells (Slominski et al. 2005c). These findings were further confirmed by the detection of NAS and melatonin using tandem liquid chromatography/mass spectrometry (LC/MS) in epidermal cells (Slominski et al. 2002a, c) and hair follicles (Kobayashi et al.

2005). These findings showed that human skin, in addition to the pineal gland and retina, possesses the intrinsic capability to synthesize melatonin (Abe et al. 1999; Carrillo-Vico et al. 2004; Finocchiaro et al. 1991; Itoh et al. 1999; Scarparo et al. 2000). Importantly, this cutaneous melatoninergic pathway operates in a compartment-specific manner since it is localized mainly to the epidermal, adnexal, and dermal cell populations (Fig. 2.5) (Slominski et al. 2008a).

Similar capability to produce melatonin was demonstrated in rodent skin (Slominski et al. 2005c). For example, in hamster skin fragments maintained ex vivo serotonin was transformed into melatonin with NAS as the intermediate product (Slominski et al. 1996b, 2002a). This transformation was time- and dose-dependent, and was stimulated by forskolin—indicating involvement of cAMP signal in this process (Slominski et al. 1996b). These findings have been confirmed in follow-up studies (Slominski et al. 2005c). Specifically, biochemical assays in mouse, rat, and hamster skin clearly demonstrated that skin of all of these species can transform serotonin to NAS, the obligatory precursor for melatonin (Semak et al. 2004; Slominski et al. 2002a, 2003b, c). Additionally, murine skin in organ culture and mouse vibrissae hair follicles can produce melatonin and its synthesis was enhanced by the addition of norepinephrine (Kobayashi et al. 2005). Interestingly, detailed analysis with bisubstrate Cole inhibitor in combination with molecular analyses showed that in rodent skin NAS production was initiated by both AANAT and NAT (Semak et al. 2004; Slominski et al. 2002a), while in C57BL6 mouse NAS appeared to be only produced by NAT (Slominski et al. 2003b, c). This latter finding provides mechanistic explanation for melatonin production in C57BL/ 6 mice at selected extracranial sites, which would require HIOMT expression since NAS produced via AANAT-independent pathways could serve as substrate for HIOMT-mediated transformation into melatonin (Ma et al. 2008; Scarparo et al. 2000; Slominski et al. 2003b, c). Our enzymatic studies excluded corporal skin of the C57BL/6 mouse in vivo as a site of melatonin production, although we detected low levels of HIOMT activity in mouse ear (Slominski et al. 2003b, c).

3.2 Melatonin Degradation

Melatonin can be degraded via indolic and kynuric pathways. The first one involves 6-hydroxylation by CYP1A1, CYP1A2, or CYP1B1 to 6-hydroxymelatonin (predominantly in the liver), which after sulfatation or glucuronidation is excreted in urine (Kopin et al. 1961; Ma et al. 2005, 2008). In the liver, the intrinsic clearance for melatonin hydroxylation by high- and low-affinity components indicated that both mitochondrial and microsomal cytochrome P450s metabolize melatonin principally by 6-hydroxylation, with O-demethylation representing minor metabolism (Ma et al. 2005). In addition, melatonin deacetylase produces 5-methoxytryptamine that is oxidized by monoamine oxidase to form 5-methoxyindoleacetaldehyde, which is converted to 5-methoxyindole acetic acid by aldehyde dehydrogenase or to 5-methoxytryptophol by alcohol dehydrogenase (Cahill and Besharse 1989; Grace et al. 1991). In the kynuric pathway, melatonin can be converted either enzymatically or non-enzymatically to N1-acetyl-N2-formyl-5-methoxykynuramine (AFMK), in a process that encompasses generation of 3-hydroxymelatonin, 2-hydroxymelatonin, melatonin 2-indolinone, 3-hydroxymelatonin, 2-indolinone, and melatonin dioxetane as intermediate products (Hardeland et al. 2009; Hirata et al. 1974; Reiter et al. 2007). AFMK synthesis involves enzymes or pseudoenzymes such as cytochrome c, horseradish peroxidase, indoleamine dioxygenase, myeloperoxidase, oxoferryl hemoglobin, or hemin as well as nonenzymatic pathway that may be activated in the presence of reactive oxygen species (ROS) or UVB (Fischer et al. 2006a; Hardeland et al. 2009; Kanda and Watanabe 2007; Seever and Hardeland 2008; Semak et al. 2005, 2008). In addition, catalase, arylamine formamidase, hemoperoxidase, and ROS can stimulate the conversion of AFMK to AMK (Hardeland et al. 2009; Kanda and Watanabe 2007; Reiter et al. 2007). Melatonin can also be demethylated to NAS by CYP2C19 or CYP1A2 (Semak et al. 2008). However, according to some authors, AFMK and AMK pathways of melatonin metabolism are insignificant at the systemic level in mouse (Ma et al. 2008).

Melatonin metabolites 5-methoxytryptamine (5-MTT) and 5-methoxytryptophol (5-MTOL) have been detected in cultured mammalian skin fragments and melanoma cells (Slominski et al. 1996b, 2002b, c), indicating similarity in the degradative pathways of melatonin metabolism in frog skin and retina (Cahill and Besharse 1989; Grace et al. 1991), including the activity of monoamine oxidase (MAO) in mammalian skin (Semak et al. 2004; Slominski 2005). It was shown that cutaneous degradation of melatonin may also include pathways known to be operative in the liver and kidney (Grace et al. 1991; Kanda and Watanabe 2007; Pandi-Perumal et al. 2006) with 6-hydroxymelatonin production as an intermediate (Fischer et al. 2006a). This shows that indolic degradative pathway is operating in the skin (Fischer et al. 2006a; Slominski et al. 1996b). However, experiments with cultured human immortalized keratinocytes have shown that melatonin is mostly metabolized to 2-hydroxymelatonin and AFMK by the kynuric pathway or through direct nonenzymatic action of UVB (Fischer et al. 2006a). Interestingly, UVB also induces AFMK utilization by keratinocytes, suggesting the involvement of arylamine formamidase in the further metabolism of AFMK to AMK (Fischer et al. 2006a). Based on the above, together with the known mechanism for melatonin degradation or transformation in peripheral organs, we proposed that in the skin melatonin can be metabolized via alternative pathways including nonenzymatic reactions (Fig. 3.1). These would exhibit species-, site-, tissue-, and cell compartment- as well as cell type-dependent differences, subjected to further modulation by environmental factors including UVR (Fischer et al. 2008b; Slominski et al. 2008a). Pathways' activities and the nature of the final product would be linked to the spatial distribution of melatonin in the skin, to the specific cell type and subcellular compartments, where the biological activity of melatonin would either be attenuated due to its degradation or be amplified by the generation of even more potent metabolites, such as AFMK or AMK (Fischer et al. 2008b; Slominski et al. 2008a).

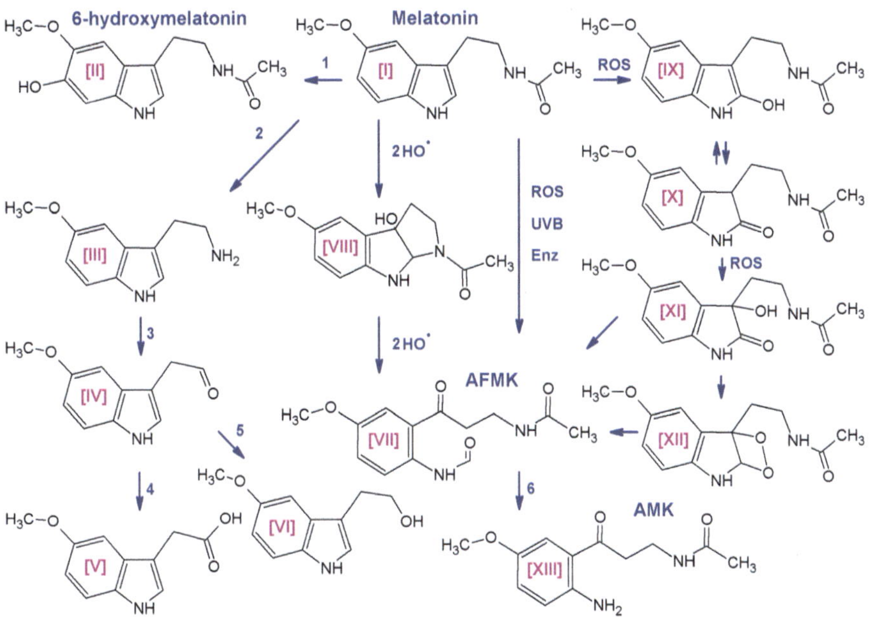

Fig. 3.1 Pathways of melatonin degradation. The indolic pathway involves 6-hydroxylation of melatonin **[I]** by CYP1A1, CYP1A2, or CYP1B1 (**1**) to 6-hydroxymelatonin **[II]**. Melatonin deacetylase (**2**) produces 5-methoxytryptamine **[III]** that is oxidized by monoamine oxidase (**3**) to form 5-methoxyindoleacetaldehyde **[IV]**, which is converted to 5-methoxyindole acetic acid **[V]** by aldehyde dehydrogenase (**4**) or to 5-methoxytryptophol **[VI]** by alcohol dehydrogenase (**5**). In the kynuric pathway, melatonin can be converted non-enzymatically to N1-acetyl-N2-formyl-5-methoxykynuramine (AFMK) **[VII]** in the process that may include generation of 3-hydroxymelatonin **[VIII]**, 2-hydroxymelatonin **[IX]**, melatonin 2-indolinone **[X]**, 3-hydroxymelatonin 2-indolinone **[XI]**, and melatonin dioxetane **[XII]** as intermediates. Enzymes or pseudoenzymes (Enz) are involved in melatonin conversion to AFMK. Different pathways can lead to the conversion of AFMK **[VII]** to AMK **[XIII]** (**6**). Reproduced with permission from the publisher (Slominski et al. 2008a)

3.3   Biological Activity of Melatonin

At the central level, melatonin functions as chronobiotic regulator (Zeitgeber; circadian pacemaker) regulating photoperiod-dependent reproduction and other biological rhythms, as well as a prominent sleep promoter (reviewed by Pandi-Perumal et al. 2006; Reiter 1991; Watson 1994; Yu and Reiter 1993). It plays a role in reproduction and sexual maturation, energy expenditure, and body mass regulation, acting via central or peripheral receptors (Hardeland et al. 2011; Pandi-Perumal et al. 2006; Reiter 1991; Watson 1994; Yu and Reiter 1993). Melatonin can additionally affect brain and immune, GI, cardiovascular, renal, bone, and endocrine functions. It shows also oncostatic, antiaging, and cell-protective activities (Bartsch et al. 2002; Bubenik 2002; Hardeland et al. 2011; Jung and Ahmad 2006; Luchetti et al. 2010; Pandi-Perumal et al. 2006; Reiter 1991; Watson 1994; Yu and Reiter 1993). Thus, melatonin

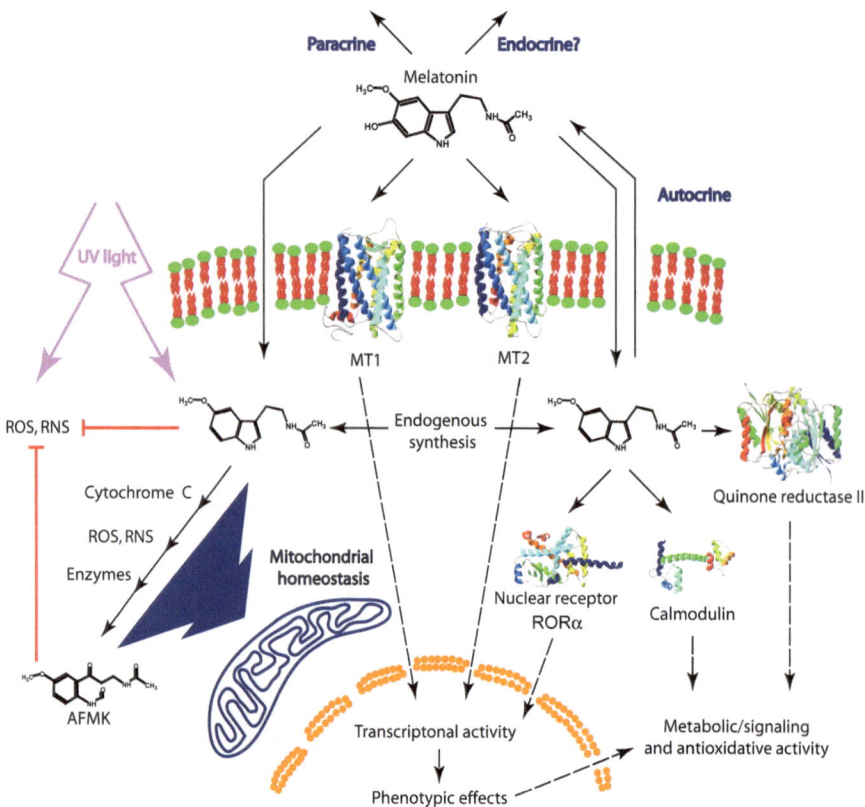

Fig. 3.2 Phenotypic effects of melatonin in the skin. Exogenous or endogenously synthesized melatonin can regulate skin cell phenotype via interaction with melatonin receptors. Non-receptor actions are mediated by interactions with intracellular proteins such as NQO2 or calmodulin. Melatonin and its metabolites can act as direct scavengers of reactive oxygen and nitrogen species (ROS and RNS) and affect mitochondrial functions. Direct effects are shown by *solid lines* and multiple reactions/signaling are shown by *dashed arrow lines*. Reproduced from (Slominski et al. 2008a) with permission from the publisher

has pleiotropic bioactivities acting as a neurotransmitter, hormone, cytokine, and biological response modifier. These functions are mediated by interactions with high-affinity receptors (Dubocovich et al. 2003; Hardeland et al. 2011; Pandi-Perumal et al. 2006; Reiter 1991; Yu and Reiter 1993). The proposed actions of melatonin or its metabolites in the skin are shown in Fig. 3.2.

Melatonin exerts its effects by interacting with specific receptors that are widely distributed throughout the body, and which are differentially expressed in various organs and tissues (Dubocovich et al. 2003, 2010). The cell surface receptor family comprises MT1 (MTNRa) and MT2 (MTNRb) whose coding region genomic structure is similar and constitutes two exons and one intron which show 60% homology at the amino acid level (Dubocovich and Markowska 2005). There are alternatively spliced isoforms of *MT* genes whose functions remain to be defined

(Slominski et al. 2003a, 2005a). It was shown that different genetic variants of *MT2* can affect body glucose homeostasis (Bouatia-Naji et al. 2009; Prokopenko et al. 2009). Both receptors belong to the family of seven-transmembrane receptors coupled to G proteins, i.e., either Gi or Gq/11—depending on the receptor type (Dubocovich et al. 2003, 2010). Melatonin activates its receptors at nanomolar or lower concentrations. Binding of melatonin to its receptors affects many cellular signaling pathways. The inhibition of cAMP and cGMP production is followed by the inhibition of CREB, PKA, and c-FOS activities, inhibition of calcium and potassium signaling, modification of protein kinase C activity, stimulation of arachidonic acid release, modification of inositol phosphate turnover, and phosphorylation of the mitogen-activated protein and extracellular signal-regulated kinases 1 and 2 (MEK1 and MEK2) as well as c-Jun N-terminal kinase (JNK) (Capsoni et al. 1994; Dubocovich et al. 2003, 2010).

The oligomerization and dimerization of MT1 and MT2 receptors appear to play a role in the regulation of cell activity (Ayoub et al. 2002; Jockers et al. 2008; Maurice et al. 2010). In addition, melatonin-related receptor (MRR or GPR50), which has high protein sequence homology with MT1 and MT2, may form heterodimers with MT1 (Ayoub et al. 2002; Jockers et al. 2008). Retinoic acid orphan receptor type α (RORα) was proposed to serve as a putative melatonin nuclear receptor (Carlberg et al. 1994; Wiesenberg et al. 1998). However, crystallographic studies have shown that RORα is in fact a cholesterol sulfate, and not melatonin, receptor (Kallen et al. 2004). Accordingly, others have shown that after binding to MT1 melatonin can indirectly regulate phenotypic activity of RORα (Hill et al. 2009). It has to be noted that there are at least 4 splicing variants of this nuclear receptor, i.e., *RORα1*, *RORα2*, *RORα3*, *RZRα* (*RORα4*) (Becker-Andre et al. 1994; Carlberg et al. 1994; Pozo et al. 2004).

Melatonin exerts also receptor-independent activities. These include broad-spectrum direct antioxidant activity or indirect actions resulting from the activation of anti-oxidative and cytoprotective pathways, a property shared by AFMK and other melatonin metabolites (Hardeland et al. 2009; Reiter et al. 2007; Tan et al. 2001). These properties define melatonin and its metabolites as anti-apoptotic and anti-mutagenic agents (Fischer et al. 2008b; Hardeland et al. 2011; Reiter et al. 2007; Slominski et al. 2008a). Melatonin can also regulate cell metabolism by acting on mitochondria (Hardeland et al. 2011; Semak et al. 2005). Some receptor-independent melatonin actions may be partly mediated via the cytosolic flavoprotein quinone reductase II (NQO2), which is involved in cellular resistance to oxidative stress and detoxification and possesses a melatonin-binding site (previously proposed as a melatonin receptor type 3 (MT3) (Jockers et al. 2008; Nosjean et al. 2000). Melatonin metabolites, including AFMK, and AMK, generated by UVB or oxidative stress can be stronger antioxidants than melatonin itself (Fischer et al. 2006a; Hardeland et al. 2007; Seever and Hardeland 2008). These receptor-independent protective actions of melatonin and its metabolites would require high intracellular levels of the molecules, which can only be met by melatonin in situ production in the relevant tissue, since cellular melatonin uptake is very limited because only 0.1% of extracellular melatonin can enter the cell (Fischer et al. 2006a).

3.4   Melatonin Receptors in the Skin

The major compartments of the skin, i.e., the epidermis, dermis, and adnexa, are targets for melatonin regulation (Fischer et al. 2008a; Slominski et al. 2005a, 2008a). More specifically, melatonin was implicated in the regulation of hair growth cycle (Fischer et al. 2008a; Kobayashi et al. 2005; Slominski et al. 2005c), cutaneous pigmentation (Slominski et al. 2004c), as well as skin physiology and pathology (Slominski et al. 2005a) including melanoma (Fischer et al. 2006c; Slominski and Pruski 1993; Yu and Reiter 1993) and vitiligo (Schallreuter et al. 2008a; Slominski et al. 1989). Since those actions have been recently discussed (Dubocovich et al. 2010) the description below will be short. The field of hair growth was the subject of an intensive research in Australia and New Zealand, where experiments on fur-covered animals revealed that melatonin stimulates hair growth. For instance, melatonin-supplemented diet increased the rate of hair growth in springtime (Welch et al. 1990). The results were confirmed in other experimental models (Ibraheem et al. 1994; Nixon et al. 1993). It is likely that fur growth is mediated by melatonin-binding sites/receptors, since these are expressed in rodent skin (Kobayashi et al. 2005; Slominski et al. 1994, 2004b). Clinical studies in women suffering from androgenic alopecia showed a positive effect of melatonin on human hair growth, suggesting it as a potential hair growth regulator in humans (Fischer et al. 2004).

Expression of cell surface MT receptors in the skin varies in different species. To illustrate, skin from the C57BL/6 mouse expresses the *MT2* gene predominantly (Kobayashi et al. 2005) or exclusively (Slominski et al. 2004b), while human skin expresses both MT receptors although a bias toward *MT1* gene expression was observed. The MT1 represents the predominant form of melatonin receptor found in both whole skin and cultured cells (Slominski et al. 2003a, 2005a). As shown by immunocytochemical studies, the expression of MT1 and MT2 proteins in human skin was cell type- and cell compartment-dependent (Slominski et al. 2005a), which suggests that the selectivity of melatonin action could be achieved by spatial compartmentalization and the specificity of signal transduction pathways. The expression of MT receptors was modified by environmental factors in a cell type-specific fashion (Slominski et al. 2003d, 2005a). For example, UVB at $100 \, mJ/cm^2$ induced expression of the *MT1* gene in neonatal epidermal melanocytes, and downregulated it in mela noma cells. *MT1* gene expression was also dependent on the donor, e.g., in some samples of dermal fibroblasts it was not detectable (Slominski et al. 2003d, 2005a). *MT2* gene expression was also upregulated or modified by UVB in epidermal keratinocytes (normal and immortalized), epidermal melanocytes, and dermal fibroblasts. UVB also induced alternatively spliced *MT2b* isoform in epidermal melanocytes, keratinocytes, and dermal fibroblasts (Slominski et al. 2003a). *MT2b* isoforms have two open reading frames (*orf*) encoding the putative proteins. The first *orf*, called MT2b1, would generate a truncated protein of 79 amino acids containing the N-terminal and first transmembrane sequence followed by 8 amino acids of MT2 with the sequence GEHHS added due to a frame shift and the addition of a stop codon. The second *orf*, MT2b2, if translated, would code a protein of 247 amino acid protein, lacking the TM 1-3 domains (Slominski et al. 2003a, 2005c).

Concerning *RORα* nuclear receptor gene, all of the tested human skin cell types (keratinocyte, melanocyte, and fibroblast lineages) expressed at least one of the three *RORα* isoforms except for *RORα3* gene (Fischer et al. 2006c; Slominski et al. 2005a). Furthermore, UVB downregulated the expression of *RORα* in HaCaT keratinocytes and upregulated it in normal neonatal melanocytes. *RORα1* and *RORα2* expression was detected only in dermal fibroblasts and in immortalized melanocytes (*RORα2* only) while being undetectable in normal epidermal melanocytes and in keratinocytes (Slominski et al. 2005a). It is likely that *RORα* gene, due to the alternative splicing (the gene contains 23 exons within a genomic region that spans 732,840 bp), codes other isoforms, as we detected additional DNA fragments of an unexpected length using standard *RORα1* and *RORα2* primers (Slominski et al. 2005a). The hair cycle-dependent expression of RORα was detected in the mouse skin (Kobayashi et al. 2005). The current challenge is to define whether the RORs still can serve as low-affinity receptors for melatonin taking into consideration that RORα is a specific receptor for sterols (Kallen et al. 2004) and that its role as a receptor for melatonin was questioned (Dai et al. 2001).

3.5 Melatonin Protects Against Skin Damage

Serving as an antioxidant and radical scavenger (Tan et al. 2002), melatonin acts as a protecting factor against UVR-induced damage in the skin (reviewed by Fischer et al. 2008b; Slominski et al. 2005a, 2008a). In fact, melatonin is able to prevent sun damage but only when it is administered prior to the UVR exposure and/or is present at the irradiation site (Bangha et al. 1996, 1997; Dreher et al. 1998, 1999). In vitro, melatonin increased cell viability in UV-irradiated fibroblasts (Lee et al. 2003), and decreased apoptosis (Ryoo et al. 2001). Melatonin also protected human leukocytes against UV-induced damage, significantly suppressing ROS formation. It had even stronger radical-scavenging properties than vitamin C and Trolox (Fischer et al. 2002). In human epidermal keratinocytes, melatonin protected against UV-induced reduction of cell viability (Fischer et al. 2006b, 2008c). This effect was found to be receptor-independent (it required high doses of melatonin), and involved anti-apoptotic activities. The interactions between pathways stimulated by UVB and melatonin may be complex since melatonin also attenuated the expression of several genes whose expression was known to be upregulated by UVB (Pisarchik et al. 2004). Thus, melatonin could have a clinically relevant protective action against UVR when used as a sun protective cream component. Although melatonin photostability is a limiting factor, its metabolites 6-hydroxymelatonin and N1-acetyl-N2-formyl-5-methoxyky-nurenamine can retain significant antioxidant activity (Maharaj et al. 2002). A challenging question is whether protective functions of melatonin depend partly on its regulation of NQO2 function, since NQO2 gene expression is ubiquitous in skin cells (Slominski et al. 2005a).

3.6   Conclusions

Melatonin is generated and metabolized in the skin to affect its phenotype as well as to serve as a protective agent against UV radiation. Some of the melatonin effects are mediated through its interaction with melatonin receptors. Other actions result from direct, receptor-independent effects of this free-radical-scavenging molecule as well as metabolic and protective effects induced by melatonin or its metabolites. The pleiotropic activities of the cutaneous melatoninergic system are mediated by cell-specific intra-, auto-, or paracrine mechanisms, allowing a counteraction or attenuation of both environmental and endogenous stressors leading to the mainte-nance of skin integrity, and perhaps affecting body's homeostasis (Fig. 3.2). Local melatoninergic systems could also modify the activities of the cutaneous neuroen-docrine network and influence global homeostasis as shown at Figs. 1.1 and 1.2.

Chapter 4
Cutaneous Cholinergic System

4.1 An Overview

Acetylcholine acts via nicotinic or muscarinic receptors. There are several subtypes of nicotinic receptors that are built of pentamers of at least 17 ($\alpha1$–$\alpha10$, $\beta1$–$\beta4$, γ, δ, ϵ) subunits (Wu and Lukas 2011). Several subunits of the same type may be present in any given receptor type (e.g., $\alpha1$ and $\alpha5$). Nicotinic receptors signal by forming ligand-regulated cation channels. There are five subtypes of muscarinic receptors (M1–M5) that act through G-protein-coupled signaling (Graef et al. 2011). Acetylcholine is synthesized by choline acetyltransferase from acetyl coenzyme A and choline. Acetylcholinesterase degrades acetylcholine to acetate and choline. Human keratinocytes synthesize, secrete, and degrade acetylcholine. Choline acetyltransferase is present in all layers of human epidermis, while acetylcholinesterase is present only in basal keratinocytes (Grando et al. 1993). The role of cholinergic system in the skin has been reviewed extensively by Grando and coworkers (2006). The role of acetylcholine-cholinergic receptor system in the skin, which among others regulates the function of eccrine glands, is well known (Fitzpatrick et al. 1993). However, the exclusive role of muscarinic system in sweat glands was challenged by finding nicotinic receptors in myoepithelial and acinar cells of those glands (Kurzen and Schallreuter 2004).

Human keratinocytes express the $\alpha3$-, $\alpha5$-, $\alpha7$-, $\alpha10$-, $\beta1$-, $\beta2$-, and $\beta4$-nicotinic receptor subunits and all types of muscarinic receptors (Grando 1997, 2006; Grando et al. 1995, 1996, 2006). The expression of these receptors changes during the process of keratinocyte differentiation. Basal keratinocytes respond to acetylcholine predominantly via nicotinic receptor $\alpha3\beta2(\beta4)$ with or without $\alpha5$ subunit and the M_2- and M_3-muscarinic receptors. Keratinocytes of the prickle layer have more $\alpha5$-containing $\alpha3$-nicotinic receptors, and also express $\alpha9$-nicotinic as well as M_4- and M_5-muscarinic receptors. $\alpha7$-nicotinic and M_1-muscarinic receptors are mainly found on keratinocytes of the granular layer of epidermis (Grando et al. 2006). The $\alpha7$-nicotinic receptor has the most prominent role in keratinocyte differentiation since its deactivation leads to the apoptosis of keratinocytes and the inhibition of their differentiation. The α3β2 receptor regulates chemokinesis of leukocytes (Arredondo et al. 2002; Chernyavsky et al. 2004a), while the activation of nicotinic receptors stimulates keratinocyte motility, with α9 subtype of nicotinic receptor being the most significant in this respect (Nguyen et al. 2000). It was suggested, that, acting simultaneously, nicotinic (primarily α7) and muscarinic (primarily M1) receptors are responsible for directional migration of keratinocytes via the Ras/Raf-1/MEK1/ERK pathway (Chernyavsky et al. 2004a; Grando et al. 2006). Also, the activation of M3-muscarinic receptors favors the expression of migratory integrins and that of M4 promotes sedentary integrins, thereby further solidifying a pivotal role of the cholinergic system in keratinocyte migration and wound re-epithelialization (Chernyavsky et al. 2004b). Muscarinic receptors' activation increased relative amounts of Ki-67, PCNA, and p53 mRNAs as well as PCNA, cyclin D1, p21, and p53 proteins affecting cell cycle (Arredondo et al. 2003).

Acetylcholine's potential role in the pathogenesis of pemphigus can be demonstrated by the fact that cholinergic receptors' activation on keratinocytes altered the expression of desmoglein 1, desmoglein 3, and the phosphorylation status of desmoglein 3 (Nguyen et al. 2003). Cholinergic agonists inhibit the antibody-induced acantholysis and adhesion molecules' phosphorylation in pemphigus vulgaris. M1 ligand binding leads to the activation of both serine/threonine and tyrosine phosphatases, whereas binding of a ligand to the α7-nicotinic receptor activates the tyrosine phosphatase and inhibits Src (v-src sarcoma (Schmidt-Ruppin A-2) viral oncogene homologue). These processes lead to the dephosphorylation of adhesion molecules and, thus, the inhibition of acantholysis (Chernyavsky et al. 2008).

SLURP family proteins regulate the function of the cholinergic system and their abnormalities are found in one of the palmoplantar keratodermas (Mal de Meleda) and psoriasis (Fischer et al. 2001; Tsuji et al. 2003). Local acetylcholine levels are increased in atopic dermatitis (Wessler et al. 2003).

Human melanocytes express the M_1–M_5 subtypes of muscarinic receptors and α1, α3, α5, α7, β1, β2, γ, and δ subunits of nicotinic receptors (Buchli et al. 2001). Acetylcholine induces pigmentation via nicotine receptors and inhibits it via M2- and M4-muscarinic receptors (Grando et al. 2006; Kurzen and Schallreuter 2004; Wallstrom et al. 1999). The activity of acetylcholine esterase is decreased in vitiligo (Iyengar 1989). The M4-muscarinic receptor may have a pivotal role in the regulation of murine hair pigmentation (Hasse et al. 2007).

All the elements of the cholinergic system are expressed in Langerhans cells and lymphocytes. Muscarinic M3 receptor expression on lymphocytes is much stronger than other receptor subtypes (Tayebati et al. 2002).

4.2 Conclusions

The cholinergic system plays a pivotal role in the regulation of keratinocytes' homeostasis. Their differentiation, motility, adhesion, and cell cycle are modified by acetylcholine. The differential expression of the receptors has been documented in detail. The α7-nicotinic receptor plays a key role in keratinocyte differentiation and α9-nicotinic, M3-, and M4-muscarinic receptors in keratinocyte migration. In the melanocytes, the activation of nicotinic receptors induces pigmentation, while the opposite is true for M2- and M4-muscarinic receptors. The cholinergic system is implicated in skin pathologies such as palmoplantar keratoderma (Mal de Meleda type), psoriasis, atopic dermatitis, vitiligo, and pemphigus. In addition, communication between the cutaneous neuroendocrine system and the rest of the body is partly achieved via the cholinergic system (Figs. 1.1 and 1.2).

Chapter 5
Corticotropin Signaling System in the Skin

5.1 CRF and Urocortins

Corticotropin-Releasing Factor (CRF), a 41 amino acid long hypothalamic neuropeptide discovered by Vale and Rivier (Spiess et al. 1981; Vale et al. 1981), and related urocortin (Urc1-3) are brain neuropeptides that regulate behavioral, autonomic, endocrine, reproductive, metabolic, and immune functions (Grammatopoulos and Chrousos 2002; Hillhouse et al. 2002; Perrin and Vale 1999). In peripheral tissues, they act as local immunomodulators with predominantly proinflammatory actions (Hasse et al. 2007; Slominski 2003b; Slominski et al. 2006c; Theoharides and Cochrane 2004) as well as they directly regulate cardiovascular, gastrointestinal, reproductive, and gestational activities (Hillhouse et al. 2002). These neuropeptides exert their regulatory activities via interaction with CRF receptors, CRF1 and CRF2, which were cloned and initially characterized by Vale's group and others (Grammatopoulos and Chrousos 2002; Hillhouse and Grammatopoulos 2006; Hillhouse et al. 2002; Perrin and Vale 1999; Slominski et al. 2001).

5.2 Expression and Functions of CRF and Urocortins in the Skin

The CRF signaling system regulates human skin homeostasis (Janjetovic et al. 2009; Slominski and Wortsman 2000; Slominski et al. 2000c, 2006c; Zmijewski and Slominski 2009b, 2010) (Fig. 5.1). In fact, we were the first to detect CRF and Urc1 production in the skin (Roloff et al. 1998; Slominski et al. 1996a, 1998b, 1999c, 2000b, c), which was stimulated by UVR and cAMP, and was inhibited by dexamethasone (Slominski et al. 1996a, 1998b). We also identified and characterized the CRF receptors in the skin and defined their functional activity (Slominski et al. 1999c, 2000c, 2001, 2006c), findings confirmed and extended by others. CRF and urocortins can inhibit proliferation of cultured human keratinocytes and melanocytes (Quevedo

A.T. Slominski et al., *Sensing the Environment: Regulation of Local and Global Homeostasis by the Skin's Neuroendocrine System*, Advances in Anatomy, Embryology and Cell Biology 212, DOI 10.1007/978-3-642-19683-6_5, © Springer-Verlag Berlin Heidelberg 2012

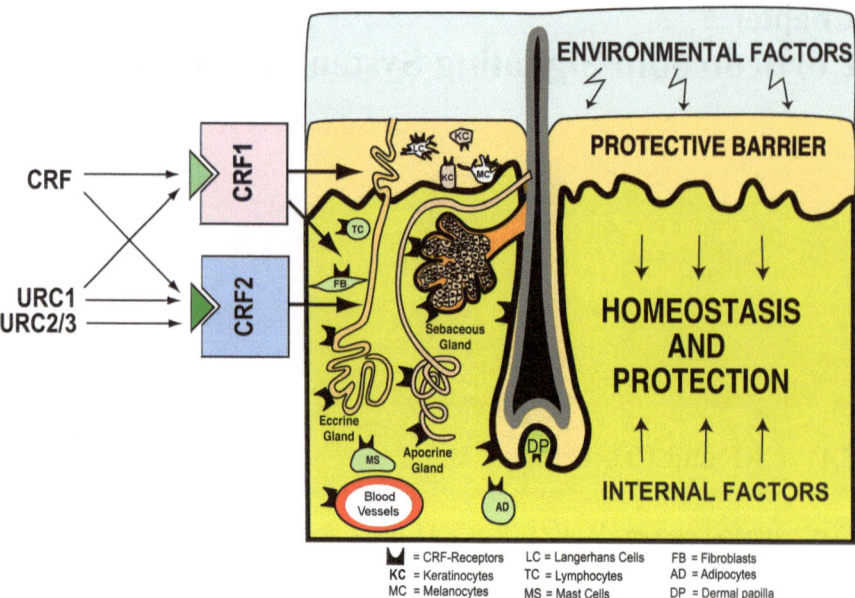

Fig. 5.1 CRF-related signaling in the skin regulates its protective and homeostatic functions. The specificity of the effect is defined either by a local production of molecules (CRF, Urc1, Urc2, or Urc3) or the type of the receptor expressed (CRF1 vs. CRF2). Reprinted from (Slominski 2009b) with permission from the publisher

et al. 2001; Slominski et al. 2000b, 2006a; Zbytek et al. 2005). They also stimulate cell differentiation (Zbytek et al. 2005) and modulate the expression of cell surface adhesion molecules and cytokine production by human keratinocytes (Quevedo et al. 2001; Zbytek et al. 2002, 2004). CRF and related peptides stimulate POMC expression and corticosterone and cortisol production (Cirillo and Prime 2011; Hannen et al. 2011; Ito et al. 2005; Rousseau et al. 2007; Skobowiat et al. 2011; Slominski et al. 2005d, e, 2006c; Vukelic et al. 2011; Zbytek et al. 2006b). These phenotypic effects of CRF in the skin are mediated by interaction with CRF1, which is the predominant receptor type expressed in human epidermis, and CRF2, and appear to be secondary to the modulation of intracellular concentrations of cAMP, IP3, Ca^{2+}, or NF-κB activity (Fazal et al. 1998; Slominski et al. 1999c, 2005e, 2006a; Wiesner et al. 2003). Moreover, CRF also stimulates steroidogenic activities in sebocytes which express both CRF1 and CRF2 (Zouboulis et al. 2002). Thus, CRF and Urc exhibit non-endocrine activities like regulation of cell proliferation, differentiation, and immune cell interactions, thereby defining these peptides as a novel type of growth factors/pleiotropic cytokines (Kauser et al. 2006; Slominski et al. 2006a, c; Zbytek and Slominski 2007; Zbytek et al. 2006a).

These pleiotropic activities of CRF and urocortins give the mechanism of regulating CRF signaling in the skin a significant priority (Slominski et al. 1999c, 2006c; Zmijewski and Slominski 2010). CRF1 is expressed in all major cellular lineages of the skin, while CRF1α prevails in human epidermis, and CRF2 is

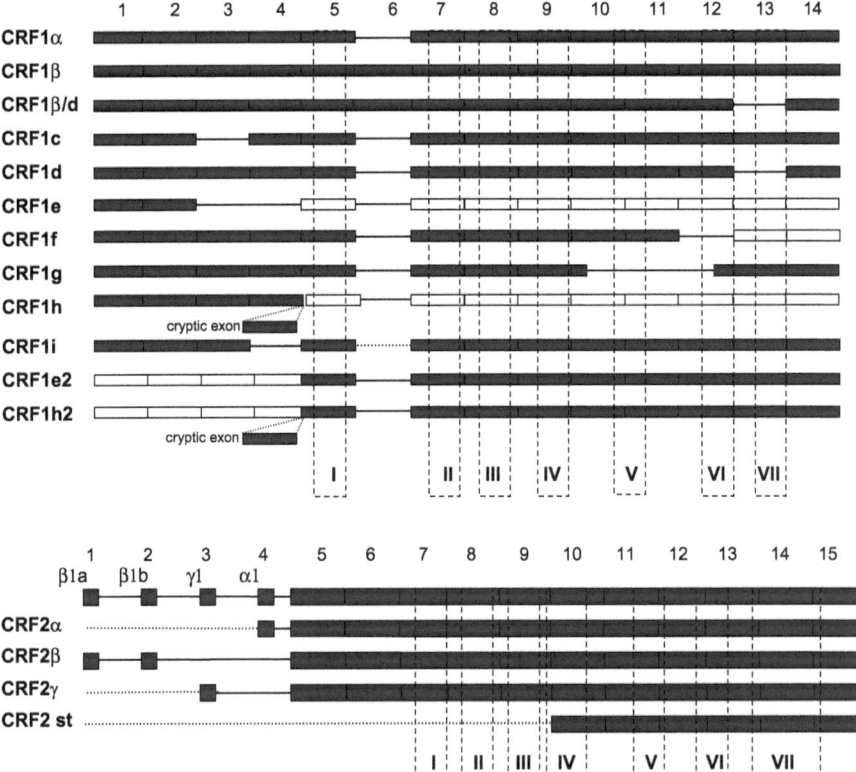

Fig. 5.2 CRF1 and CFR2 receptors and their alternative splicing variants. *Upper panel*: Human CRF1 gene consists of 14 exons and due to alternative splicing at least ten isoforms can be generated with seven found in human skin (Pisarchik and Slominski 2001; Slominski et al. 2006c). Coding exons are shown in blue and noncoding exons due to frame shift followed by in-frame premature stop codon are showed as white squares. *Lower panel*: CRF2 gene contains 15 exons and at least three alternative transcription start codons. Due to alternative splicing at least three main isoforms can be created (CRF1α, β, γ) and four additional isoforms could be synthesized from full-length mRNA by employing alternative start codons (CRF2α1, β1a, β1b, γ1 as shown on the top of the panel). Exon numbers are marked on the top of each panel. Transmembrane segments of the 7TM domains are shown as *squares* (*dashed line* with number I–VII) (Slominski et al. 2006c; Zmijewski and Slominski 2010)

expressed in all mouse cutaneous compartments (Slominski et al. 2001, 2004a, 2006c). A major challenge in this area is the functional implication of the coupling of different CRF1 isoforms to different signal transduction systems (Figs. 5.2 and 5.3) (Janjetovic et al. 2009; Pisarchik and Slominski 2001, 2004; Slominski et al. 2001, 2004a, 2006a, c; Zmijewski and Slominski 2009a, b, 2010). This differential coupling could provide the mechanistic explanation for observed organ- and cell type-dependent variability in the phenotypic response to CRF as previously suggested (Pisarchik and Slominski 2001, 2004; Slominski et al. 2004a, 2006c; Zmijewski and Slominski 2010).

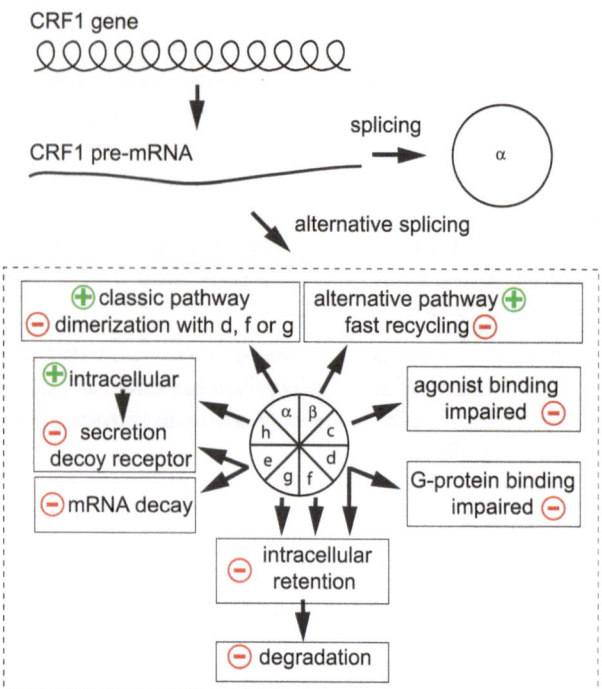

Fig. 5.3 Regulation of the CRF signaling by CRF1 isoforms. *CRF1* gene contains 14 exons and only one isoform of the CRF1β receptor (also called pro-CRF1) is coded by all exons. CRF1 transcript is also subjected to alternative splicing resulting in at least eight isoforms. Recent studies showed that the expression and/or co-expression of CRF1 isoforms is responsible for the modulation of CRF1 signaling mediated by main CRF1α or alternative CRF1β isoform. Soluble isoforms (CRF1e and h) were also found to stimulate CRF or modify Urc signaling when co-expressed with CRF1α. "Minus" sign indicates inhibition of CRF signaling on different levels including: fast mRNA decay (CRF1e), dimerization, and subsequent intercellular retention resulting, most probably, in premature receptor degradation (CRF1α with CRF1d, CRFf, or CRFg), decoy receptor mechanism (CRF1h and e when secreted), agonist binding impairment (CRF1c), or G-protein binding inhibition (CRF1d). For details, see (Zmijewski and Slominski 2010). Reproduced with permission from the publisher

5.3   Splicing of CRF Receptor Transcripts

Alternative splicing of mRNA is one of the most important mechanisms accountable for genomic variability in higher eukaryotes (Luco et al. 2011). Therefore most human genes, including those coding for G-protein-coupled receptors (GPCRs), are sources for multiple protein isoforms.

CRF receptors are members of class B (secretin family) of G-protein-coupled receptors (GPCRs) and are closely related to calcitonin, growth-hormone-releasing hormone (GHRH), glucagon, glucagon-like peptides, parathyroid hormone (PTH), pituitary adenylate-cyclase-activating peptide (PACAP), and secretin receptors

(Lagerstrom and Schioth 2008). Members of the class B of the GPCR receptor family bind to peptides longer than 27 amino acid residues and are expressed in the majority of endocrine and non-endocrine cells (Hillhouse and Grammatopoulos 2006; Perrin and Vale 1999; Slominski and Wortsman 2000).

In humans, the *CRF1* gene, which contains 14 coding exons, was mapped to chromosome 17 (17q12-q22) (Polymeropoulos et al. 1995). CRF2 gene has 15 exons (Hillhouse and Grammatopoulos 2006; Slominski et al. 2001) and is located on chromosome 7 (7p14.3). The coding sequences of the two receptors show high degree of homology, although not uniformly distributed along the sequence. A comparison of the protein sequences revealed three distinct regions of homology, corresponding to the structural domains of CRF receptors. Extracellular domain (ECD) of the CRF receptor is responsible for substrate recognition and binding, and this region showed the lowest homology (40%) between CRF1 and CRF2. This feature most probably reflects differential affinity to ligands (CRF and Urc 1-3) (Hillhouse and Grammatopoulos 2006). On the other hand, 7TM domain and intracellular and extracellular loops are highly conserved with homology of around 80%. The most conserved part of the CRF receptor is the third intracellular loop involved in the interaction with G-proteins (Hemley et al. 2007; Hillhouse et al. 2002). Despite a high level of similarity, the pattern of splicing variants of the two known CRF receptor genes (*CRF1 and CRF2*) seems to be unique for each pre-mRNA (Fig. 5.2).

5.3.1 CRF1 mRNA Splicing Variants

Processing of pre-mRNAs encoding *CRF* receptors may result in an alternative splicing with at least ten variants of *CRF1* mRNA (α, β, c, d, e, f, g, h, and i) (Hillhouse and Grammatopoulos 2006; Karteris et al. 2011; Pisarchik and Slominski 2001, 2002; Slominski et al. 2006c, 2007a; Zmijewski and Slominski 2010). Importantly, all of those isoforms except newly discovered *CRF1β/d* and *CRF1i* were found in human skin (Mikhailova et al. 2007; Slominski et al. 2007a; Zmijewski and Slominski 2010, 2011). Interestingly, only one isoform of *CRF1* (CRF1β) contains all 14 exons, while the main functional isoform, CRF1α, has an exon 6 spliced out. The exon 6 seems to be unique for CRF1β, although a recent study by Karteris and coworkers (2011) revealed a new isoform, named CRF1β/d since it shared the properties of isoform CRF1β (exon 6) and CRF1d (lack of exon 13). This finding raises a theoretical possibility of the expression of all *CRF1* splicing variants with and without exon 6. Other *CRF1* isoforms might be divided into three groups: soluble receptors (CRF1e and CRFh), receptors with defects in the ECD (isoform c), and receptors with impaired 7TM domain (CRF1d, f, g, and also CRF1β/d fits to this group). The detailed exonal organization of CRF1 isoforms is shown in Fig. 5.2 and was discussed elsewhere (Hillhouse and Grammatopoulos 2006; Slominski et al. 2006c; Zmijewski and Slominski 2010, 2011). It has to be noted that alterative splicing of *CRF1* also results in a frame shift which introduces premature stop codon to the sequence of CR1e, f, g, and h. Alternative splicing of CRF1 receptor mRNA seems to be a conserved phenomenon in evolution because CRF1 isoforms were also identified in rat (Hillhouse and Grammatopoulos 2006; Slominski et al. 2001), mouse (Pisarchik and Slominski 2001), and hamster (Pisarchik and Slominski 2002). Also, some splice variants are conserved among the members of the family B of GPCRs. For instance, characteristic deletion of exon 13 was found in *CRF1* isoform d and calcitonin receptor (Grammatopoulos et al. 1999; Markovic et al. 2008; Seck et al. 2005; Zmijewski and Slominski 2009b). Theoretically, it is possible that, due to the alternative splicing, also "headless" CRF1 receptor isoforms could be coded by CRF1 pre-mRNA (Slominski et al. 2006c). Their presence was predicted based on the known mRNA sequences of isoforms CRF1e and h (Pisarchik and Slominski 2001; Zmijewski and Slominski 2010). The mRNAs of those isoforms—due to the presence of the alternative code premature stop codon and alternative ATG start cordons—could theoretically allow for the synthesis of "headless" isoforms of CRF1 receptor. Such ECD domain-missing isoforms were identified for the closely related calcitonin receptor (Nag et al. 2007). However, proof for the existence of CRF1 "headless" isoforms remains to be provided.

5.3.2 CRF2 Splicing Variants

In theory, *CRF2* gene has a capacity for similar number of isoforms as shown for CRF1 (http://www.ncbi.nlm.nih.gov/IEB/Research/Acembly/), although only CRF2α, β, γ, and soluble sCRF2 isoforms of *CRF2* were well characterized (Grammatopoulos and Chrousos 2002; Hillhouse et al. 2002). In contrast to *CRF1* gene, the gene of *CRF2* receptor has at least three alternative starting codons coded by alternative exons located on 5'-end of the *CRHR2* gene (see Fig. 5.2). In addition, the presence of the soluble isoform of *CRF2* (sCRFR2α) was also reported in mouse brain (Perrin et al. 2003) and headless isoform of *CRF2* was found in stomach (GenBank accession No. E12750; Patent: JP199707289-A).

5.3.3 Expression of CRF1 and CRF2 Isoforms in the Skin

In human skin, *CRF1* gene is expressed in both epidermal and dermal compartments, whereas *CRF2* is detected predominantly in adnexal structures such as hair follicles (Kauser et al. 2006; Slominski et al. 2004a) or sebaceous glands (Kauser et al. 2006). It seems that this pattern is characteristic for humans because mouse skin expresses both *CRF1* and *CRF2* (Slominski et al. 2004a, 2007a), and both of them take part in the regulation of skin physiology (Kauser et al. 2006; Slominski et al. 2004a, 2006c).

All studied human epidermal and dermal cell lines express the main CRF1 isoform α and for some cells like adult epidermal keratinocytes and melanocytes, it is the only isoform found under normal conditions (Pisarchik and Slominski 2001; Slominski et al. 2001). Also, melanocytes, keratinocytes, and fibroblasts found in hair follicles express CRF1α (besides the previously mentioned CRF2) (Kauser et al. 2006; Slominski et al. 2004a, 2006c). However, neonatal epidermal keratinocytes, dermal fibroblasts, and several melanoma cell lines express multiple CRF1 variants (Pisarchik and Slominski 2001; Slominski et al. 2004a).

5.3.4 Modulation of the Expression of CRF1 Isoforms and Its Physiological Relevance

Although the regulation of the alternative splicing of CRF receptor genes' remains unknown, a theoretical model of alternative splicing with a potential involvement of U1 and U2 small nuclear ribonucleoproteins (snRNPs), splicing activators, and Ser/Thr kinases was proposed (Markovic and Grammatopoulos 2009). Here, we will discuss only biological factors which affect *CRF* receptor splicing with relevance to human skin.

The *CRF1* expression pattern and alternative splicing is regulated by diverse physiological and pathological factors, including cell growth conditions or exposure to the ultraviolet irradiation (Zmijewski and Slominski 2010). In human immortalized HaCaT keratinocytes, CRF1α is the only isoform expressed in confluent culture. However, fast growing (subconfluent) cells express multiple isoforms including α, c, and e (Zmijewski and Slominski 2009b). In addition, the expression of CRF1 mRNA and protein increases with confluence of HaCaT keratinocyte cultures (Zmijewski and Slominski 2009b). The above phenomena may also explain differences in the CRF1 expression between neonatal and adult epidermal keratinocytes (Slominski et al. 2004a, 2007a).

The pathological conditions can influence the expression of CRF1 receptor as shown in skin biopsies from psoriatic patients (Tagen et al. 2007; Zmijewski and Slominski 2009b). It is worth mentioning that most of the studied melanomas expressed multiple CRF1 isoforms including CRF1α, except for SKMEL-188 melanoma cells that exclusively expressed CRF1d (Pisarchik and Slominski 2001; Slominski et al. 2004a). This raises a question whether CRF1 splicing is involved in the pathogenesis of skin hyperproliferative (malignant or benign) and inflammatory diseases and whether external and internal stressors can affect skin physiology through context-dependent *CRF1* alternative splicing leading to the differential CRF signaling in this organ (Slominski 2009b; Slominski et al. 2006c; Zmijewski and Slominski 2010).

A single-nucleotide polymorphism (SNP) of *CRF1* gene might be an additional factor with high impact on *CRF1* expression and splicing. There is a growing body of evidence that SNPs of *CRF1* are associated with several human pathologies including hypertension, abusive behavior, and depression (Kamdem et al. 2008; Schmid et al. 2009; Wasserman et al. 2008). Therefore, it is possible that SNPs might influence the *CRF1* gene expression and/or splicing of its pre-mRNA.

Recent studies revealed a potential mechanism and significance of CRF1 isoforms' expression in the epidermal and dermal cell lines and other models (Jin et al. 2007; Karteris et al. 2011; Markovic et al. 2008; Pisarchik and Slominski 2001, 2002; Slominski et al. 2006a, c, 2007a; Sztainberg et al. 2009; Zmijewski et al. 2007; Zmijewski and Slominski 2009b). As stated above, the pattern of CRF1 isoform expression depends on cell type and culture growth condition. These data also correlate with the observed changes in the responsiveness to CRF or urocortin. The current model of the regulation of CRF signaling confirms the central role of CRF1α and suggests at least modulatory roles for the other CRF1 isoforms in signal transduction (Fig. 5.3) (Zmijewski and Slominski 2010). For example, alternative splicing may decrease the levels of CRF1α transcripts. Isoform CRF1e is a good candidate. This isoform consists of only first coding exons followed by the premature termination codon in exon 5, resulting in a frame shift due to the removal of exons 3 and 4. This isoform could be subjected to fast decay mechanisms. An additional mechanism of the regulation of CRF1 function by the expression of multiple splicing variants may include heterooligomerization of CRF1 isoforms leading to changes in CRF1α trafficking to the cell membrane, its localization, and function. Oligomerization of GPCR is a well-known mechanism of the regulation and activation of the GPCR A family. Interestingly, oligomerization of CRF1 isoforms was found in different cell compartments of HaCaT epidermal keratinocytes (Zmijewski and Slominski 2009a) and also in pituitary AtT-20 cells (Zmijewski and Slominski 2009b). These findings confirmed earlier studies which showed dimerization of CRF1α (Kraetke et al. 2005) and heterodimerization of CRF1 with vasopressin V1b receptor (Young et al. 2007). Thus, the co-expression of CRF1α with CRF isoforms defective in 7TM domain (d, f, g) may result in the retention of heterodimers inside the cells and thus inhibition of the translocation of the newly synthesized CRF1α receptors to the cell membrane. Based on structural modeling of CRF1 isoforms, it was concluded that alterations to the sequence caused by alternative splicing should result in the instability of receptors in the cell membrane (Slominski et al. 2006c). Indeed, CRF1d, when overexpressed in HaCaT keratinocytes, localized predominantly to the endoplasmic reticulum, and CRF1f and CRF1g co-localized within Golgi cisterns. Thus, none of the CRF1 isoforms with impaired 7TM domain showed proper membrane localization (Zmijewski and Slominski 2009a, b). The overexpression of those isoforms can influence downstream signaling including cAMP, IP$_3$ production, and calcium mobilization, followed by altered transcriptional activity (Grammatopoulos and Chrousos 2002; Grammatopoulos et al. 1999; Hillhouse and Grammatopoulos 2006; Markovic et al. 2008; Pisarchik and Slominski 2004; Slominski et al. 2006a, 2007b; Wietfeld et al. 2004; Zmijewski et al. 2007; Zmijewski and Slominski 2009a, b, 2010). The third mechanism of the regulation of CRF signaling may involve direct alterations of the receptor function. CRF1d, f, and g isoforms and their heterodimers with CRF1α, found predominantly inside the cell, could not

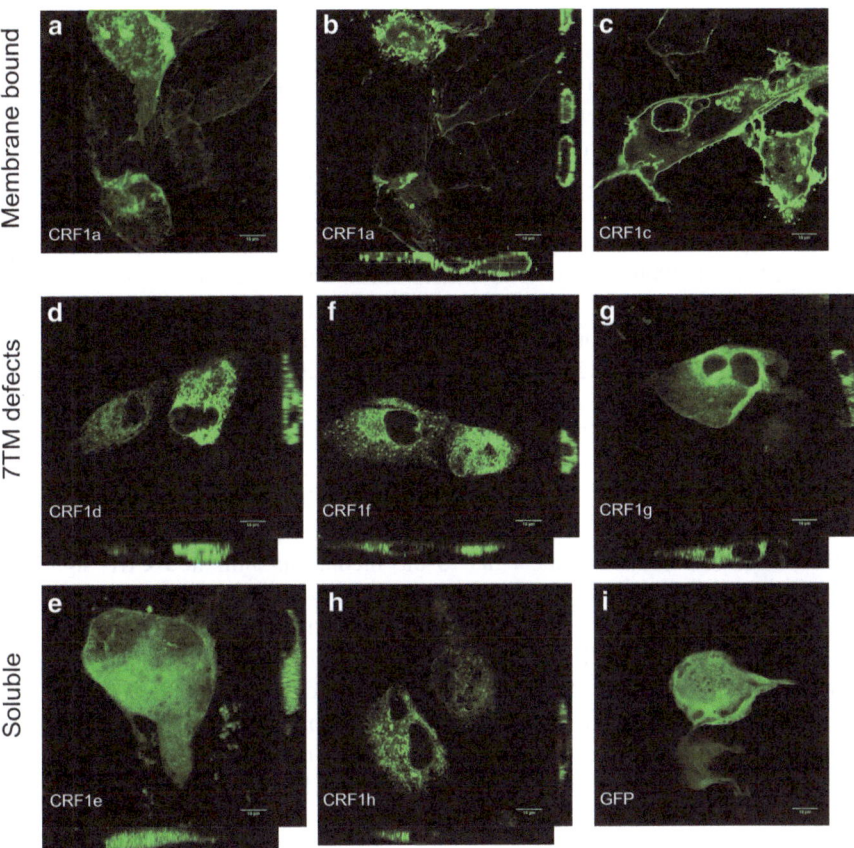

Fig. 5.4 Localization of the CRF1 isoform tagged with GFP in human adult ARPE-19 cells. Adult retinal pigment epithelium cells (ARPE-19) as alternative to the melanocyte model of pigment-producing cells showed similar intracellular distribution of the CRF1 isoform to that described previously in HaCaT keratinocytes (Zmijewski and Slominski 2009b) and ATT-20 pituitary cells (Zmijewski and Slominski 2009b). Isoforms CRF1α (Panels **a** and **b**) and CRF1c (Panel **c**) with full-length 7TM are found predominantly within cell membrane. CRF1 isoforms with defects (CRF1d—Panel **d**, CRF1f—Panel **f**, CRF1g—Panel **g**) within 7TM region show intracellular localization. The soluble isoforms (CRF1e—Panel **e**, CRF1h—Panel **h**) are localized predominantly inside the cells. The isoform CRF1e (Panel **e**) is the only isoform found inside the cell nucleus (similarly as GFP alone—Panel **i**). ARPE-19 cells were transfected with constructs carrying CRF1 isoforms fused with GFP (Zmijewski and Slominski 2009c) and images (as Z stacks) were collected with Zeiss LSM 510 laser scanning microscope (Zeiss, Germany). On the bottom and right sides of Panels **b**, **d**, **f**, **g**, **e**, and **h** cross sections (from Z stacks) were shown to emphasize three-dimensional localization of the CRF isoforms. On Panel **a**, Z stack projection (average intensity) of APRE-19 cells overexpressing CRF1α is shown to emphasize the presence of this isoform on the cell surface. The controls are represented by ARPE cells transfected with GFP alone (Panel **i**)

be activated by extracellular ligands (Pisarchik and Slominski 2004; Slominski et al. 2007b; Zmijewski and Slominski 2009a, 2010). Even if isoforms with impaired 7TM domains (CRF1d, f, and g) would reach proper cell membrane localization, the downstream signaling would be impaired due to improper binding/activation of G-protein (Zmijewski and Slominski 2009a, b, 2010). On the other hand, the isoform CRF1c, despite its proper membrane localization (Slominski et al. 2006c; Zmijewski and Slominski 2009a, b), has a deletion of exon 3 encoding the main part of receptor's ECD, which resulted in impaired ligand binding (Slominski et al. 2006c; Zmijewski and Slominski 2009a). Indeed, CRF1 expressed in COS-1 cells failed to bind antagonist ($[I^{125}]$oCFR (Ross et al. 1994). An increase in cAMP production was only observed after stimulation with high concentration of human CRF indicating attenuation of signaling by CRF1c in comparison to CRFα (Karteris et al. 1998).

A fourth type of the CRF signaling modulation is represented by the soluble CRF1h isoform (Slominski et al. 2006c; Zmijewski and Slominski 2011). The intracellular retention and co-localization within ER of CRF1h were shown in cells overexpressing this isoform (Zmijewski and Slominski 2009a, b). In addition, CRF1h was released from the cells to the media and inhibited CRF signaling (Pisarchik and Slominski 2004; Zmijewski and Slominski 2009b). If this process occurs in vivo, it would be consistent with a well-known mechanism of action of soluble receptors (also called decoy receptors) and similar to the function of CRF2 soluble isoform (sCRF2) (Chen et al. 2002). The described models of the CRF signaling regulation via the expression of CRF1 isoforms may be not unique to skin cells but can represent a global mechanism valid for other tissues and cell lineages, because many organs and tissues express multiple CRF1 isoforms (Fig. 5.4).

5.3.5 Conclusions

CRF, urocortins, and CRF receptors are widely expressed in human skin. The phenotype of skin cells is affected by endogenously produced CRF and urocortins via a variety of signal transduction systems. Furthermore, expression of multiple CRF1 isoforms may represent additional means of regulating CRF-mediated stress responses at different levels (Fig. 5.2). Different stress signals were found to influence *CRF1* splicing and signaling, but there are also indications that other factors such as small nucleotide polymorphism may influence CRF signaling by modification of the CRF1 pre-mRNA splicing. The recently proposed mechanism of the CRF signaling regulation by its receptor splicing may explain the observed changes in cell responsiveness to CRF and CRF-like ligands (Fig. 5.3).

Modulation of *CRF1* splicing may have a direct impact on the differentiation and proliferation of various skin cell types. Alterations of *CRF1* splicing mechanism can ultimately lead to the development or aggravation of symptoms of several skin pathologies including acne, psoriasis, or skin cancer (Fig. 5.1).

Chapter 6
Steroidogenesis in the Skin

6.1 An Overview

Adrenocortical steroidogenesis is initiated by the interaction of ACTH with melanocortin receptor type 2 (MC2-R) that stimulates secretion and production of cortisol via the activation of steroidogenic enzymes and cholesterol mobilization or transport into mitochondria (Felig and Frohman 2001; Payne and Hales 2004). Prolonged ACTH effects include induction of corresponding enzymes, and the entire process is linked to the ACTH-stimulated cAMP production (Felig and Frohman 2001; Payne and Hales 2004). The biochemical pathway is initiated by the rate-limiting enzyme cytochrome P450scc (encoded by the *CYP11A1* gene) that cleaves cholesterol side chain to produce pregnenolone, precursor to all steroids (Payne and Hales 2004; Tuckey 2005). In a classical pathway, pregnenolone is further transformed to corticosterone or cortisol through sequential action of 3β-HSD/isomerase, P450c17, P450c21, and P450c11β enzymes and then released into circulation (Felig and Frohman 2001; Payne and Hales 2004). Specifically, 3β-HSD/isomerase transforms pregnenolone to progesterone, and 17(OH)pregnenolone to 17(OH)progesterone, P450c17 catalyzes 17α-hydroxylation of pregnenolone or progesterone to 17(OH)pregnenolone or 17(OH)progesterone, and P450c21 hydroxylates progesterone to deoxycorticosterone (DOC) or 17(OH)progesterone to 11-deoxycortisol, while P450c11β hydroxylates DOC to corticosterone and 11-deoxycortisol to cortisol (Felig and Frohman 2001; Payne and Hales 2004; Simard et al. 2005). ACTH also stimulates production of mineralocorticoids and sex hormones via MC2-R and has a trophic effect on the adrenal cortex (Felig and Frohman 2001) (Fig. 6.1). Low-level production of cortisol via the classical pathway was reported in peripheral tissues (Taves et al. 2011) including the gastrointestinal tract, brain (Davies and MacKenzie, 2003; Do Rego et al., 2009), immune cells (Costa et al. 2009; Vacchio et al. 1994), and also in colon cancer (Sidler et al. 2011). In addition, cortisol levels are also regulated by 11β-HSD1 (Draper and Stewart 2005), which at high NADPH/NADP$^+$ ratios transforms cortisone to cortisol (Draper and Stewart 2005; Tomlinson et al. 2004). Cortisol can also be

A.T. Slominski et al., *Sensing the Environment: Regulation of Local and Global Homeostasis by the Skin's Neuroendocrine System*, Advances in Anatomy, Embryology and Cell Biology 212, DOI 10.1007/978-3-642-19683-6_6, © Springer-Verlag Berlin Heidelberg 2012

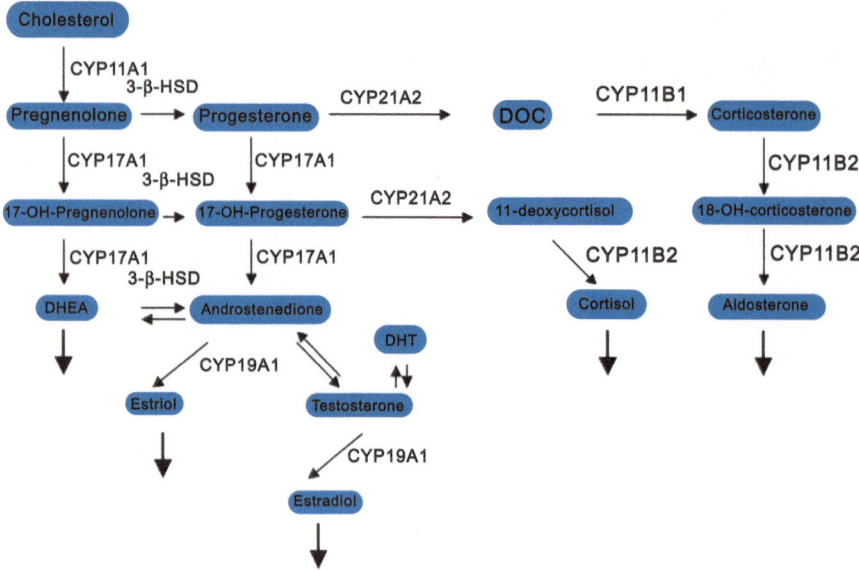

Fig. 6.1 A scheme of the steroidogenic pathway

transformed to cortisone by NADP$^+$-dependent enzyme 11β-HSD2 that acts exclusively as a dehydrogenase (Tomlinson et al. 2004). The expression of both 11β-HSD1 and 11β-HSD2 enzymes was shown in placenta, kidney, liver, fibroblasts, and adipocytes (Bujalska et al. 1997, 2002; Ricketts et al. 1998; Tiganescu et al. 2011).

6.2   Cutaneous Corticosteroidogenic System

We were the first to demonstrate that human skin expresses genes encoding enzymes involved in the sequential metabolism of cholesterol to pregnenolone and to corticosteroids including cytochromes *P450scc*, *P450c17*, and *P450c21*, and the *MC2-R* (receptor for ACTH) genes (Slominski et al. 1996d). These findings were later complemented by the demonstration of these enzymes' functional activity in epidermal and dermal skin cells (Rogoff et al. 2001; Slominski et al. 1999b, 2000a, 2002d, 2004a) and that cutaneous steroidogenesis begins from cholesterol (Slominski et al. 2004d, 2007a). Rapid metabolism of progesterone and deoxycorticosterone (DOC) was shown in rodent skin and cultured human normal and malignant skin (Slominski et al. 1999b, 2000a, 2002d). Thus, production of corticosterone and DOC-like steroid species was shown in rat skin (Slominski et al. 2000a). Also cultured malignant melanocytes showed progressive transformation of progesterone to DOC, 18-hydroxy-DOC, and corticosterone, but not to aldosterone (Slominski et al. 1999a). Cortisol and corticosterone production was further documented in normal epidermal melanocytes (Slominski et al. 2005e) and dermal fibroblasts (Slominski et al. 2005d, 2006b), with final evidence on cortisol synthesis by human skin cells provided by liquid chromatography-mass spectrometry (LC/MS) analysis (Slominski et al. 2005e, 2006b). In agreement, cortisol production was demonstrated in human hair follicles (Ito et al. 2005; Sharpley et al. 2009), and cortisol production by epidermal and dermal skin cells was later confirmed by other authors (Cirillo and Prime 2011; Hannen et al. 2011; Vukelic et al. 2011). Interestingly, cutaneous cortisol production was mediated by both CYP11B1 and 11β-HSD1 activities (Cirillo and Prime 2011; Hannen et al. 2011; Slominski et al. 2007a; Tiganescu et al. 2011; Vukelic et al. 2011). While some authors found cortisol production by early passages of human epidermal keratinocytes (Cirillo and Prime 2011; Hannen et al. 2011; Vukelic et al. 2011), we (Slominski et al. 2005e) and others (Milewich et al. 1986) did not detect cortisol in late passages of epidermal keratinocytes or melanocytes. In agreement with the last finding, in HaCaT keratinocytes, progesterone and DOC were metabolized rapidly to steroid products different from corticosterone, aldosterone, and cortisol (Slominski et al. 2002d). These discrepancies may be due to the contamination of primary cultures of keratinocytes by other cell types (e.g., melanocytes) or differences in culture conditions. Cortisol production by skin cells was regulated by ACTH and factors raising cAMP level (Slominski et al. 2005e), IL-1, as well as by wound response (Vukelic et al. 2011) and high energy ultraviolet radiation (Skobowiat et al. 2011).

6.3 Production of Sex Hormones in the Skin

The skin is an important organ transforming dehydroepiandrosterone (DHEA) and DHEA-sulfate (DHEA-S) or androstenedione, which predominantly originate from systemic circulation, to active sex hormones (Labrie et al. 2003; Ohnemus et al. 2006; Zouboulis 2004; Zouboulis and Degitz 2004; Zouboulis et al. 2007) (Fig. 6.1). In addition, in the skin local steroidogenic system (Ito et al. 2005; Slominski and Wortsman 2000; Slominski et al. 2002d, 2004d, 2005d, e, 2008b; Taves et al. 2011) produces 17(OH)pregnenolone and 17(OH)progesterone that are further metabolized to DHEA with their following metabolism to androgens and estrogens (Fig. 6.1), or other steroidal products (Slominski et al. 2002d, 2009a, c).

DHEA of systemic or local origin is transformed by 3β-HSD into 4-androstenedione, and 5-androstene-3β,17β-diol into testosterone, while 17β-HSD converts DHEA into 5-androstene-3β,17β-diol, 4-androstenedione into testosterone, and androstanedione into DHT (Labrie et al. 2000, 2003; Milewich et al. 1991; Simard et al. 1993; Zouboulis et al. 2008; Zouboulis and Degitz 2004). Cutaneous testosterone is also converted into dihydrotestosterone (DHT) by the action of a 5α-reductase (reviewed by Zouboulis et al. 2008). Skin and subcutaneous adipose tissue is also an important site of estrogen production, in particular after menopause (Labrie et al. 2003; Ohnemus et al. 2006; Zouboulis et al. 2007).

Furthermore, it is an important site of estrogen and androgen activation (Fig. 6.1) (Labrie et al. 2000; Ohnemus et al. 2006; Zouboulis et al. 2007; Zouboulis and Degitz 2004). The locally produced sex hormones modify skin phenotype and function via interactions with the corresponding androgen and estrogen receptors (Labrie et al. 2003; Ohnemus et al. 2006; Randall et al. 1993; Slominski and Wortsman 2000; Zouboulis 2004; Zouboulis et al. 2007; Zouboulis and Degitz 2004).

6.4   Conclusions

Mammalian skin is an extra-adrenal site of mineralo/glucocorticoid synthesis, which can be regulated by endogenous and environmental factors (Slominski et al. 2007a, 2008a). Furthermore, the skin is an important site for estrogen and androgen production, activation, or metabolism (Labrie et al. 2003; Ohnemus et al. 2006; Zouboulis et al. 2007). Interestingly, skin production of steroids seems to be cell type dependent and subjected to the regulation by external factors such as ultraviolet radiation. These steroids act in intra-, auto-, or paracrine fashions to regulate local homeostasis. Moreover, skin and its subcutaneous tissue constitute an important source of estrogens and androgens in females, especially after menopause.

Chapter 7
Equivalent of Hypothalamo–Pituitary–Adrenal Axis in the Skin

7.1 Systemic HPA Axis

The work of Hans Selye was fundamental in defining the hypothalamic–pituitary–adrenal (HPA) axis as the body's important coordinator of responses to systemic stress (Selye 1936; Seyle 1976). The HPA functional structure has been completed by determining that hypothalamic corticotropin-releasing factor (CRF) acts as the regulator of the production of ACTH and β-endorphin in the anterior pituitary (Spiess et al. 1981; Vale et al. 1981). The HPA pathway (Fig. 7.1) is triggered by various stress factors which activate production of CRF in the paraventricular nucleus (PVN) (Chrousos 1995; Chrousos and Gold 1992; Owens and Nemeroff 1991). In pituitary CRF binds to CRF type 1 receptors (CRF1) (Aguilera et al. 2001; Hillhouse and Grammatopoulos 2006; Perrin and Vale 1999) increasing production and secretion of the proopiomelanocortin (POMC)-derived peptides, i.e., ACTH, MSH, and β-endorphin (Hillhouse and Grammatopoulos 2006; Pritchard and White 2007; Smith and Funder 1988). The arginine vasopressin (AVP) produced by the PVN can also act synergistically with CRF in activating the HPA axis (Chrousos 1995; Itoi et al. 2004). In the adrenal cortex ACTH, by interacting with the MC2 receptors (MC2-R), stimulates production and secretion of cortisol in humans or corticosterone in rodents. These corticosteroids counteract the effects of stressors by mobilization of energy reserves, buffering tissue damages, and suppressing immune system. Moreover, corticosteroids via feedback mechanisms inhibit the HPA axis through the suppression of CRF and POMC production. The HPA axis is also controlled by cytokines, tissue modifiers, and growth factors, which can be either produced in the brain or by peripheral tissues including cells of the immune system. Thus, there are various ways of controlling stress responses at the level of hypothalamus or pituitary that bypass the central brain coordinating centers (c.f. in Besedovsky and Rey 2007; Blalock and Smith 2007; Chesnokova and Melmed 2002) (Fig. 7.1).

A.T. Slominski et al., *Sensing the Environment: Regulation of Local and Global Homeostasis by the Skin's Neuroendocrine System*, Advances in Anatomy, Embryology and Cell Biology 212, DOI 10.1007/978-3-642-19683-6_7, © Springer-Verlag Berlin Heidelberg 2012

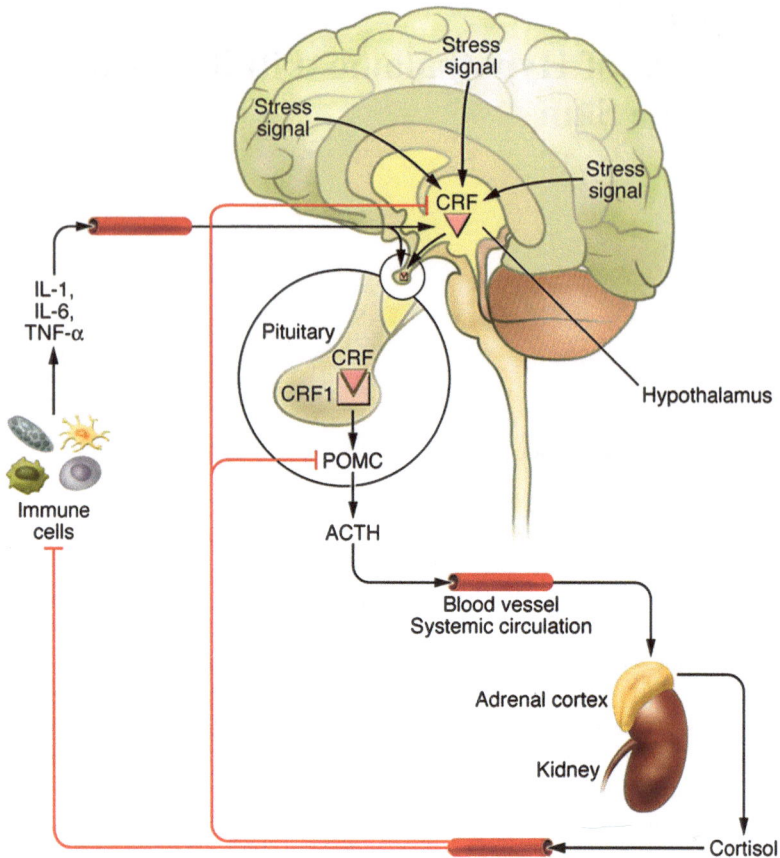

Fig. 7.1 Organization of the systemic HPA axis. Modified from Fig. 1 published in Slominski (2007)

7.2 HPA Axis Homologue Is Expressed in the Skin

More than a decade ago, we proposed that skin expresses a homologue of the HPA axis to regulate local stress responses (Slominski et al. 1996a). This concept was based on finding all molecular elements of the HPA axis in the mammalian skin, i. e., CRF, CRF1, POMC, ACTH, MC2-R, glucocorticoid receptors, and genes coding steroidogenic enzymes (Slominski 1991; Slominski et al. 1992, 1993b, c, 1995, 1996c, 2000d). Over the last 15 years, our and other laboratories provided definitive evidence that skin expresses CRF and the POMC-derived β-endorphin, ACTH and α-MSH, the corresponding CRF1, melanocortin (MC) and opiate receptors, along with the key enzymes of corticosteroid synthesis, which results in the cutaneous production of corticosterone and cortisol (Arck et al. 2006; Rogoff et al. 2001; Slominski et al. 2000c, 2004d, 2006b, 2007a; Slominski and Wortsman 2000; Tobin 2006; Tobin and Kauser 2005b). Furthermore, we presented data

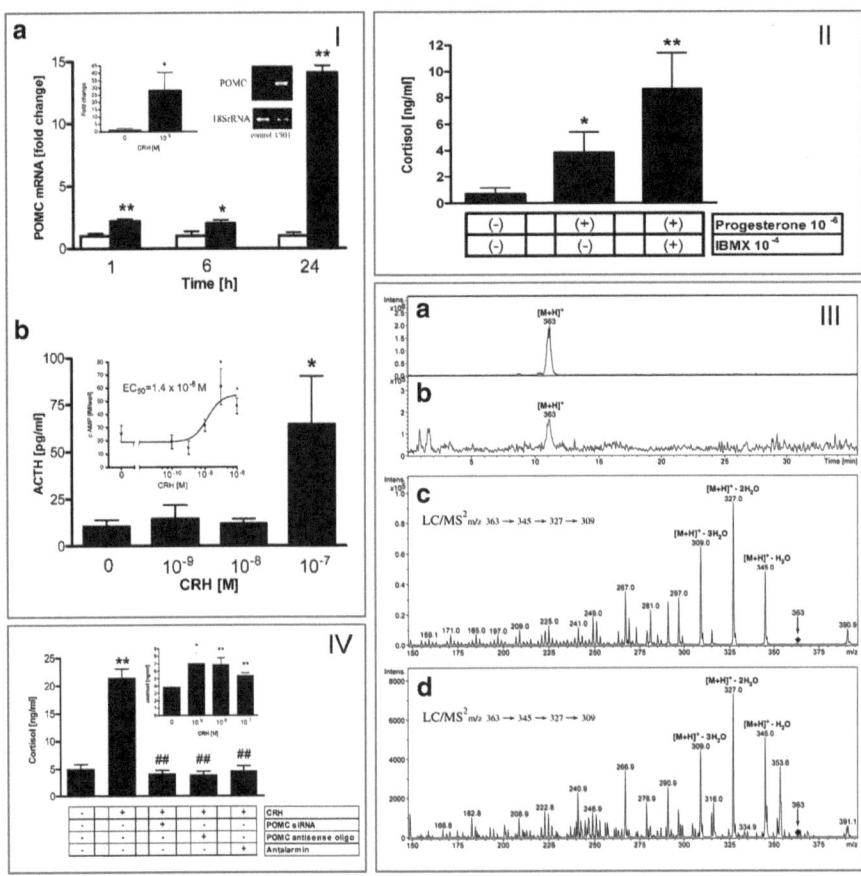

Fig. 7.2 HPA algorithm is expressed in normal human melanocytes. (**a**) CRH stimulates POMC gene expression and ACTH production with attendant stimulation of cAMP. (**b**) Cortisol production is enhanced by the addition of progesterone and/or IBMX (inhibitor of phosphodiesterase). (**c**) Cortisol is identified by LC/MS^2 in melanocytes (standard = **a, c**; conditioned media from melanocytes = **b, d**). (**d**) CRH stimulates cortisol production that is dependent on POMC expression and CRF1 signaling. Reproduced from Slominski et al. (2005e) with permission from the American Physiological Society

indicating that CRF can stimulate cortisol production in skin cells via POMC cleavage products (Fig. 7.2) (Ito et al. 2005; Slominski et al. 2005e, 2006b, 2007a). These studies provided accumulating evidence that the cutaneous stress response system follows the functional hierarchy of the central HPA with its direct local phenotypic consequences and systemic implications (Slominski et al. 2007a, 2008b; Slominski 2009b) (Fig. 7.3).

Importantly, CRF, POMC, and corresponding receptors were co-expressed in cultured keratinocytes, melanocytes, or dermal fibroblasts (Slominski et al. 2000c, 2006b) as well as their co-expression was demonstrated in vivo in the skin by in situ hybridization or immunocytochemistry (Funasaka et al. 1999; Ito et al. 2005;

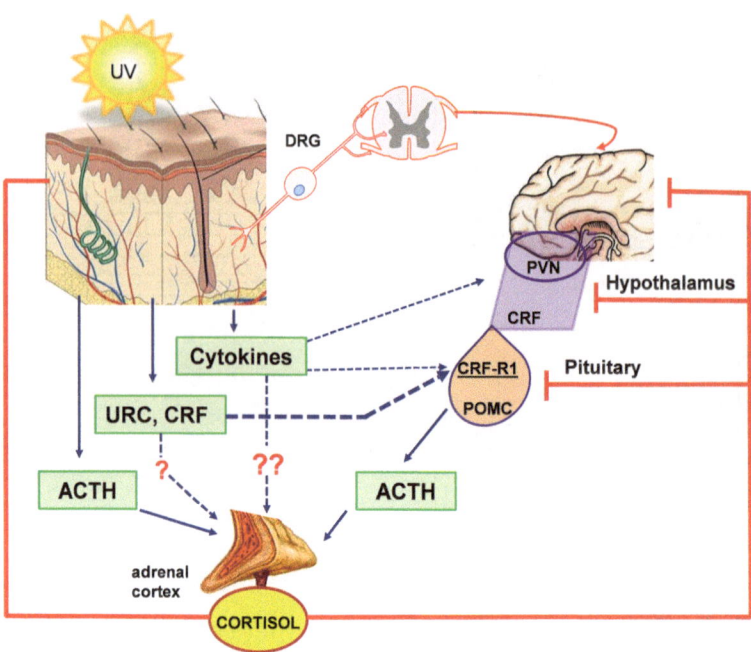

Fig. 7.3 Skin stress response system can activate the central HPA with its direct homeostatic, metabolic, and phenotypic effects. We hypothesized that global responses (at body's level) to UVR may be initiated in the skin and involve a simultaneous activation of sensory receptors and local production of humoral messages (Slominski 2005; Slominski and Wortsman 2000; Slominski et al. 2008b). These signals are either delivered by ascending nerve routes to the brain or by circulation to hypothalamus to activate CRF production in the PVN; CRF then would enter hypophyseal portal circulation with a subsequent activation of CRF1 in the pituitary. Skin humoral signals can also enter directly the pituitary from the circulation. The final outcome for these signaling processes results in the stimulation of ACTH release and of POMC activity. Alternatively, although less likely, the cutaneous factors can bypass the HPA and enter adrenal gland directly from the circulation. The net effect of all these processes is the release of cortisol/ corticosterone and induction of steroidogenesis with subsequent metabolic and homeostatic effects

Kauser et al. 2006; Kono et al. 2001; Rogoff et al. 2001; Slominski et al. 2000c, 2006b). Finally, the expression of the executive arm of the HPA, i.e., production of cortisol and corticosterone, has been clearly demonstrated in cultured epidermal keratinocytes and melanocytes as well as in dermal fibroblasts (Hannen et al. 2011; Slominski et al. 2005d, e, 2006b; Vukelic et al. 2011). Moreover, production of cortisol and corticosterone was also shown in human hair follicles maintained in vitro (Ito et al. 2005; Sharpley et al. 2009), and in rodent skin. In fact, skin contains an entire biochemical apparatus necessary to transform cholesterol to cortisol and corticosterone (Slominski et al. 2007a) including the capability to produce pregnenolone (Slominski et al. 2004d; Thiboutot et al. 2003) and its further sequential transformation to progesterone, deoxycorticosterone (DOC), 18-hydroxy-DOC, cortisol, and corticosterone (Dumont et al. 1992; Ito et al. 2005; Rogoff et al. 2001; Slominski and Wortsman 2000; Slominski et al. 1999b, 2000a,

2005d, e, 2006b). The cutaneous expression of the above HPA axis elements is nonrandom, but is organized into functional, cell type-specific regulatory loops with a structural hierarchy similar to the central HPA (Slominski et al. 2007a) (Fig. 7.3). Specifically, exogenous CRF interacted with CRF1 on cultured human epidermal melanocytes and dermal fibroblasts stimulating cAMP production with subsequent increases of POMC gene expression and production of ACTH (Slominski et al. 2005d, e, 2006b). Similarly, CRF stimulated POMC production in immortalized normal and malignant melanocytes expressing CRF1 receptor (Slominski et al. 2006b; Zbytek et al. 2006a), and CRF stimulated POMC and αMSH production by epidermal and follicular melanocytes (Rousseau et al. 2007). Most importantly, normal human melanocytes responded to CRF, ACTH, and factors raising intracellular cAMP with an increased production of cortisol and corticosterone, which was dependent on functional CRF1 receptor, since CRF1 receptor's antagonists abolished the effect, and on POMC expression, because silencing of the POMC gene abolished this effect (Fig. 7.2) (Slominski et al. 2005e). Thus, melanocytes, cells of neural crest origin, not only produce CRF, but also respond to it following an algorithm of the central HPA axis, adjusted to the local environment. While fibroblasts also responded to CRF and ACTH with enhanced production of corticosterone, cortisol levels were not regulated by axis, since cortisol was produced constitutively (Slominski et al. 2006b), thus indicating a partial departure from the classical algorithm of the HPA regulation.

The HPA components were also demonstrated in organ-cultured human scalp skin (Ito et al. 2005) where exogenous CRF increased POMC expression with sequential increases in ACTH and cortisol production. These were consistent with in situ co-localization of CRF, CRF1, and POMC in the skin and hair follicles (Kono et al. 2001; Rogoff et al. 2001; Slominski et al. 2000c). Exogenous cortisol inhibited CRF, MC2-R, and ACTH expression by interacting with glucocorticoid receptors (Ito et al. 2005) which was in agreement with earlier studies which showed inhibition of the POMC and CRF expression by dexamethasone (a synthetic glucocorticoid) in mouse skin (Ermak and Slominski 1997) and cultured human skin cells (Slominski et al. 1998b). Thus, mammalian skin expresses a fully functional HPA axis equivalent which encompasses local CRF, ACTH, and cortisol/corticosterone synthesis, and secretion with a negative feedback regulation of CRF and POMC expression mediated by glucocorticoid receptors' activation (Slominski et al. 2007a). The stimulatory role of ACTH on cortisol production by human epidermal cells has been confirmed recently (Cirillo and Prime 2011; Vukelic et al. 2011)

It is likely that the CRF-driven patterns of steroidogenic responses can be differential depending on cell subpopulation, their tissue localization, and microenvironment (Fig. 5.1) (Slominski et al. 2007a). We have proposed a crucial role of paracrine communications in the skin where keratinocytes, melanocytes, fibroblasts, immune cells, and nerve endings can serve as signal initiators (CRF, ACTH, cortisol, or corticosterone) and recipients (binding to corresponding receptors) (Slominski 2005). The latter function would include an active and compartment-specific intercellular cross-talk and exchange of intermediates of the steroidogenic pathway.

7.3   Regulation of the Cutaneous HPA Axis

To serve a role of a coordinator and executor of local responses to stress, cutaneous homologue of the HPA should be activated by specific physical, chemical, and biological skin stressors in organized fashion encompassing local production of CRF and POMC-derived peptides which interact with their respective receptors (Fig. 7.3). Indeed, an exposure of skin or skin cells to UVR stimulated in a time- and dose-dependent manner expression of CRF and POMC genes which was followed by the production and release of CRF, β-endorphin, and ACTH peptides, expression of CRF1, PC1, MC2, MC1, CYP11A1, and CYP11B1, and production of cortisol (Chakraborty et al. 1999; Pisarchik and Slominski 2001; Skobowiat et al. 2011; Slominski et al. 1996b; Zbytek et al. 2006b). The stimulatory effects were predominantly seen after exposure to UVB or UVC with only limited responses to UVA (Skobowiat et al. 2011).

In epidermal melanocytes, the UVB-induced stimulation of the *CRH* promoter was suppressed by both the inhibitors of protein kinase A (PKA) and a plasmid overexpressing dominant mutant CREB (Zbytek et al. 2006b). Accordingly, UVB stimulated CREB phosphorylation, the binding of phosphorylated CREB to CRE sites in the CRF promoter, and the activity of the reporter gene construct driven by consensus CRE sites, while the mutation in the CRE site of the *CRF* promoter rendered the corresponding reporter gene construct less responsive to UVB (Zbytek et al. 2006b). In addition, pharmacological inactivation of CRF1 by selective inhibitors abrogated the UVB-stimulated induction of POMC (Zbytek et al. 2006b). These results indicate that UVR induces CRF1 signaling by stimulating the PKA pathway with the subsequent stimulation of POMC production, which imitates HPA's organizational structure. Our most recent results also show that the ability to activate or modify the "cutaneous HPA" elements is dependent on highly energetic UV wavelengths (UVC and UVB) implying a dependence on their noxious activity (Skobowiat et al. 2011).

7.4   Functional Activity of the Cutaneous HPA

7.4.1   Local Effects

All of the HPA elements (CRF and/or POMC signaling systems and steroidogenic activities), separately or in concert, can have profound phenotypic effects in the skin (Figs. 5.1 and 7.3) (Slominski et al. 2000c, 2006c, 2007a) and may affect systemic body responses via neuroendocrine and hormonal signal transmission (Slominski 2005; Slominski et al. 2008b). In the skin, these interactions can follow the classical pathway CRF → CRF1 → POMC → ACTH → corticosterone/cortisol. However, the context-dependent departures from this central algorithm such as CRF → CRF1, CRF → CRF1 → POMC, POMC → ACTH + MSH + β-END and POMC → corticosterone/cortisol may take place. The important local pheno-typic outcomes of the entire axis or its departures are protective measures against environmental stressors (physical, biological, and chemical insults). This results from fine-tuning and selective regulation of skin pigmentation, barrier function, adaptive and innate immunity, and adnexal structures' activity. The elements of the cutaneous HPA can also counteract skin pathology, such as inflammatory and autoimmune disorders as well as hyperproliferative and dysplastic processes, in order to protect and restore skin homeostasis. In this context, non-endocrine activities of CRF and related urocortins make CRF signaling in the skin an important regulatory system. Similarly, a central role is assigned to the insult-regulated POMC expression and its context-dependent processing in the skin because the chemical nature of the final peptides defines the phenotypic effect (Slominski et al. 2000c, 2004d, 2007a). Finally, local steroidogenic activity would protect skin homeostasis and counteract pathologic processes but also, on the other hand, terminate protective responses to prevent their potential dyshomeostatic effects (Slominski 2009b; Slominski et al. 2007a, 2008b).

7.4.2 Systemic Effects

The possibility of the communication between local and systemic HPA axes as well as the differential activation of skin-derived axis elements poses an interesting research question (Slominski 2005; Slominski et al. 2008b) due to the apparent evolutionary conservation of a similar organization at both central and peripheral levels (Slominski 2007). Although all environmental factors noxious to the skin could participate in this communication (Slominski and Wortsman 2000), the role of solar radiation in illustrating the above-mentioned communication and regula-tion of body homeostasis is the most instructive (Fig. 7.3) (Slominski et al. 2008b). Namely, UVR regulation of systemic homeostasis via HPA could include stimula-tion of CRF synthesis in the hypothalamus along neutrally transmitted signals from the skin. An increased hypothalamic CRF release would automatically lead to the activation of the existing HPA with cortisol/corticosterone serving as final effector. The HPA axis could be entered at the level of hypothalamus, pituitary, or adrenal gland by skin-derived humoral messages including, respectively, cytokines (action on hypothalamus and/or pituitary), CRF/Urc 1 (action on the pituitary and, possibly, adrenal gland), or ACTH (action on the adrenal gland). This type of regulation would represent a fundamental paradigm shift in neuroendocrinology and photobiology with profound implications for clinical medicine (see above and below).

These concepts are underscored by the observation that humans and horses exposed to sunlight led to increased serum levels of α-MSH and ACTH (Holtzmann et al. 1982, 1983), while experimental whole body exposure to UVB increased β-LPH and β-endorphin serum levels (Belon 1985; Levins et al. 1983). This model may provide mechanistic explanation for a well-known phenomenon of systemic immunosuppressive action of UVB (Kripke 1994) or serve as an alternative explanation for the reported cases of the attenuation of multiple sclerosis in some patients after exposure to UVR, a phenomenon so far linked to the increased production of vitamin D (Becklund et al. 2010). Thus, our model of the UVR-mediated activation of central HPA axis may serve as a rational background for a phototherapy of systemic autoimmune disorders or other pathologies. Lastly, our model may provide mechanistic explanation of the recently described phenomenon of "UVR addiction" (Kourosh et al. 2010; Nolan et al. 2009) caused by cutaneous β-endorphin production.

7.5 Common Origin of the Central and Peripheral HPA

Evolutionary conservation of a similar HPA-like organization at central and peripheral levels has been documented (Arck et al. 2006; Slominski 2005; Slominski and Wortsman 2000; Slominski et al. 2000c, 2001, 2004d, 2006c, 2007a, 2008b). The common ectodermal origin of brain and epidermis raises the fundamental question of whether the peripheral CRF signaling system is an evolutionary duplicate of its central homologue or whether the brain itself has adopted the preexisting peripheral CRF response system during evolution of the central nervous and endocrine systems. Since it had been shown that cytokines and growth factors can modify CRF and POMC-related functions in pituitary and brain (Slominski and Wortsman 2000; Slominski et al. 2006c) and that CRF can also act as a growth factor/cytokine [a function that develops at the periphery (Kauser et al. 2006; Slominski et al. 2006a, 2006c; Zbytek and Slominski 2007)], we proposed a new hypothesis on the integumental origin of the HPA axis (Slominski 2007). We suggest that the primordial HPA (Fig. 7.4) had first developed in the integument to regulate its defensive activity against the hostile environment and pathogens. Key elements of this system include CRF-related peptide(s) that coordinate innate immune activity and skin barrier formation via CRF1 (an integrating receptor) and thus, both directly and indirectly, affect the expression of the proinflammatory cytokines such as IL-1 and TNF-α. The feedback inhibitory loop begins with CRF1-activated POMC-derived production/secretion and culminates with the production/secretion of corticosterone/cortisol that "shuts off" HPA axis activity and inhibits skin barrier activity. The intermediate signaling molecules (POMC peptides) can both weaken the skin protective barrier by their immunosuppressive action and strengthen it by stimulating melanogenesis as well as direct antimicrobial effects. Thus, the protective barrier functions could be regulated and fine-tuned by the primordial HPA, because of the close association of all of its elements. During evolution, the main algorithm CRF > CRF1 > POMC > ACTH > corticosterone/cortisol may have been adapted and perfected by the central neuroendocrine system to form the HPA axis which has separated anatomically and functionally from the immune system and the skin (Slominski 2007; Slominski et al. 2007a, 2008b). In this context, the retained cutaneous HPA may serve as an evolutionary record of the primary system (Slominski 2007) and, paradoxically, the systemic stress response can weaken the

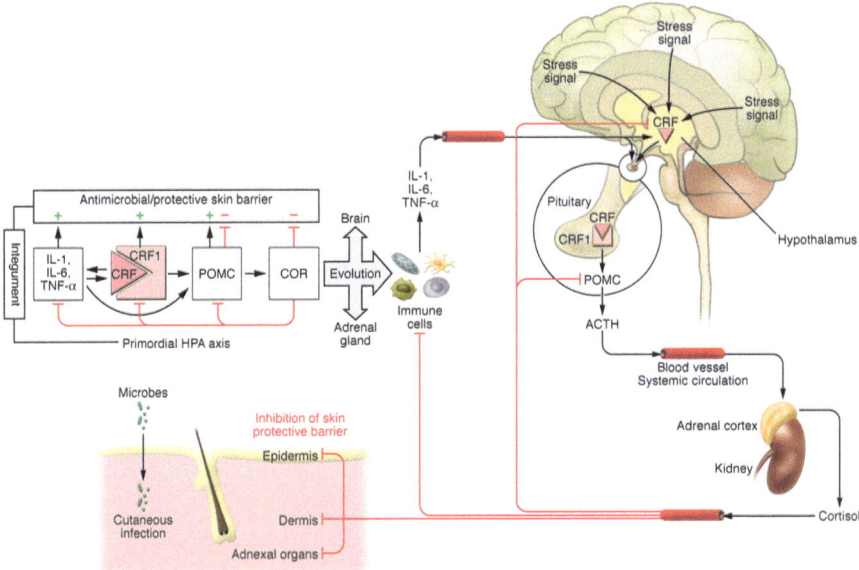

Fig. 7.4 Proposed evolution of the stress response system. Reproduced with permission from the publisher (Slominski 2007)

cutaneous antimicrobial defenses as a result of cortisol/corticosterone release from the adrenal glands (Aberg et al. 2007) (Fig. 7.4). It is also possible that both systems had derived from a common precursor and evolved in parallel, maintaining neuro-immune-endocrine communication during this process that has helped to preserve this fine organization of stress responses at the systemic and local levels.

7.6   Conclusions

Many important elements of local and systemic responses to environmental stressors (biological, chemical, and physical with predominant role of UVR) origi-nate in the skin, and they involve multiple pathways encompassing activation of different components of the cutaneous and systemic HPA. Hence, signals generated by the integrated actions of CRF, POMC peptides, and cortisol/corticosterone may counteract the local effects of the environment. Furthermore, depending on the type of the stressor and its intensity, the skin can activate systemic HPA either via neural transmission by afferent nerve fibers to the brain or by skin-derived factors which may activate pituitary gland or directly act on the adrenal cortex (Fig. 7.3).

Chapter 8
Cutaneous Secosteroidal System

8.1 7-Δ Steroids

Few years ago, it was shown that the P450scc system cleaves the side chain of 7-dehydrocholesterol (7DHC, provitamin D3) to produce 7-dehydropregnenolone (7DHP) (Guryev et al. 2003; Slominski et al. 2004d). 7DHP is a substrate for a novel metabolic pathway for the synthesis of steroids with two double bonds in 5 and 7 positions, collectively called steroidal 5,7-dienes (Fig. 8.1), which, hypothetically, may be produced in the skin since cutaneous CYP11A1 (P450scc) expression was confirmed (Slominski et al. 2004d). The role of this new pathway was supported ex vivo by demonstrating the efficient metabolic transformation of 7-dehydrocholesterol to 7DHP by adrenal glands and by mitochondria isolated from rat skin (Slominski et al. 2009c). HPLC separations, UV spectra, and mass spectrometry identified 7DHP, 22-hydroxy-7DHC, and 20,22-dihydroxy-7DHC as the major products with additional minor products defined as 17-hydroxy-7DHP and 7-dehydroprogesterone (Slominski et al. 2009c). These findings defined a novel steroidogenic pathway: 7DHC → 22(OH) 7DHC → 20,22(OH)$_2$7DHC → 7DHP, with potential further metabolism of 7DHP mediated by 3β-HSD or CYP17 along the Δ^4 and Δ^5 steroidogenic pathways, with the production of 7-dehydroprogesterone and 17(OH)7DHP as intermediates (Slominski et al. 2009c). The existence of this synthetic pathway is documented by the accumulation of pregna- and androsta-5,7-dienes and their hydroxylated derivatives in the Smith–Lemli–Opitz syndrome (SLOS), characterized by 7DHC Δ-reductase deficiency, an enzyme responsible for the conversion of 7DHC to cholesterol (Marcos et al. 2004; Shackleton et al. 1999, 2002; Tint et al. 1994). Most recently, we have found that human placenta ex utero can transform 7DHC to 7DHP and further to 7-dehydroprogesterone (Slominski et al. submitted for publication).

A.T. Slominski et al., *Sensing the Environment: Regulation of Local and Global Homeostasis by the Skin's Neuroendocrine System*, Advances in Anatomy, Embryology and Cell Biology 212, DOI 10.1007/978-3-642-19683-6_8, © Springer-Verlag Berlin Heidelberg 2012

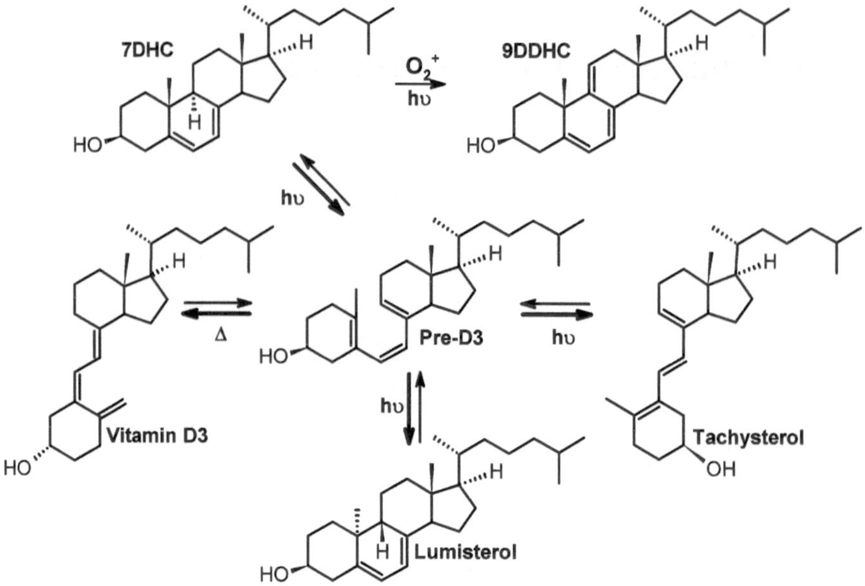

Fig. 8.1 UVB-induced production and transformation of 7-dehydrocholestrol. Reprinted from Slominski (2009a) with permission from the publisher

8.2   Secosteroidogenesis

The UVB-driven isomerization of 7DHC to vitamin D_3 [(3b,5Z,7E)-9,10-secocholesta-5,7,10(19)-trien-1a,3b,25-triol] is one of the most fundamental chemical reactions in vertebrates (Holick 2003; Holick and Clark 1978; Holick et al. 1995). This conversion is initiated by photolysis of the unsaturated B ring on absorption of UVB solar energy of 290–320 nm wavelength producing a pre-D_3 intermediate, followed by its slow isomerization to three main products: vitamin D_3, tachysterol$_3$, and lumisterol$_3$. The rate of isomerization depends on the dose of absorbed UVB energy, wavelength, temperature, and the presence of biological membranes (Fig. 8.1) (Buchli et al. 2001; Holick et al. 1995; Tian and Holick 1999). In response to high doses of UVB, another pathway of vitamin D_3 degradation/isomerization yields 5,6-transvitamin D_3, as well as suprasterols I and II were also demonstrated in the skin (Buchli et al. 2001). In biological systems provitamin D3 (7DHC) can also be transformed to 5,7,9(11)-trienes in a chemical process that involves an interplay between singlet oxygen and photosensitizers (Albro et al. 1994; Chignell et al. 2006; De Fabiani et al. 1996; Feng et al. 2006; Valencia and Kochevar 2006).

The 5,7-dienal steroids described above as well as P450scc-derived hydroxyl derivatives of 7DHC undergo UVB-induced transformation to androsta-calciferols (aD), pregna-calciferols (pD), and novel hydroxyderivatives of vitamin D_3 (Kim and Lee 2010; Li et al. 2010; Slominski et al. 2004d, 2009a, 2009c; van Beek et al. 2008; Zmijewski et al. 2011) (Fig. 8.2). Furthermore, P450scc

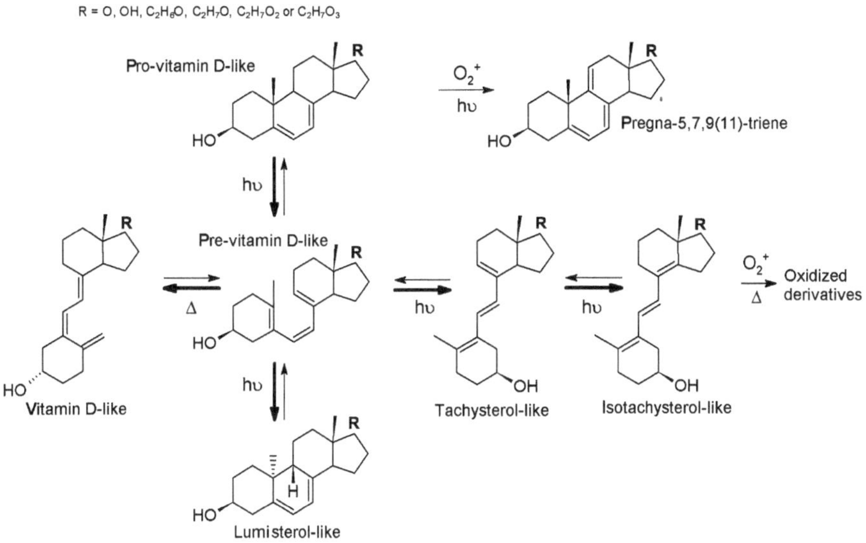

R = O, OH, C_2H_6O, C_2H_7O, $C_2H_7O_2$ or $C_2H_7O_3$

Fig. 8.2 UVB-induced transformation of pregna- or androsta-steroidal 5,7-dienes. hν - photon's energy in reaction, Reprinted from Slominski (2009a) with permission from the publisher

also metabolizes vitamin D_3 to $20(OH)D_3$, $20,23(OH)_2D_3$, $22(OH)D_3$, $20,22$ $(OH)_2D_3$, $17,20,23(OH)_3D_3$, and several other D_3-hydroxyproducts. Novel P450scc-derived secosteroids show anti-proliferatory and pro-differentiation activities in a cell type-restricted fashion that depends on the length of the side chain (Janjetovic et al. 2009, 2010; Li et al. 2010; Nguyen et al. 2009; Slominski et al. 2009a, c, 2010, 2011d; Zbytek et al. 2008; Zmijewski et al. 2009a, 2010). These compounds are anti-tumorigenic, can stimulate keratinocyte differentiation, and inhibit NF-κβ, acting by binding to vitamin D receptor (VDR) as its partial receptor agonists. They are as potent as $1,25(OH)_2D_3$, however, unlike $1,25$ $(OH)_2D_3$, only weakly stimulate CYP24 expression (Janjetovic et al. 2009, 2010; Slominski et al. 2009a; Zbytek et al. 2008). Importantly, $20(OH)D_3$ and $20,23$ $(OH)_2D_3$ did not affect calcium homeostasis at concentrations as high as 3–4 µg/kg (Slominski et al. 2010, 2011b). Thus, we discovered novel metabolic pathways initiated by the enzymatic action of cytochrome P450scc (CYP11A1) that produces biologically active novel secosteroids or their precursors, the systemic and local significance of which, including their occurrence in the skin, remains to be defined.

8.3   Vitamin D Activity in the Skin: An Overview

The role of skin in the physiology and pathology of vitamin D_3 and its derivatives was a subject of extensive reviews; therefore, below we present only short overview and refer the reader to more extensive descriptions (Bikle 2011a, b, c; Denzer et al. 2011;

Field and Newton-Bishop 2011; Holick 2003, 2008; Lehmann et al. 2004; Pinczewski and Slominski 2010; Reichrath 2007).

The epidermal keratinocytes not only are the site of the photochemical transformation of 7-dehydrocholesterol to vitamin D_3 but also possess the entire enzymatic machinery capable of activating and inactivating vitamin D_3 and its derivatives. Vitamin D_3 is activated by sequential hydroxylation in position 25 by CYP27A1 and in position 1α by CYP27B1 to form calcitriol, i.e., $1,25(OH)_2D_3$ [(1a,3b,5Z,7E)-9,10-secocholesta-5,7,10(19)-trien-1,3,25-triol)]. $1,25(OH)_2D_3$ is inactivated by the action of CYP24 yielding $1,24,25(OH)_3D_3$. The phenotypic effects of vitamin D_3 and some of its derivates are mediated by its interaction with VDR which is expressed in all skin cells including keratinocytes, melanocytes, fibroblasts, and other resident cells of the skin. VDR belongs to the family of nuclear receptors and has ligand-activated pleiotropic activities including inhibition of cell proliferation, stimulation of cell differentiation, and modulation of immune functions of skin resident and immigrant cells, to name the most important. Vitamin D is also involved in the regulation of skin barrier function, modulation of skin stress responses, regulation of hair follicle cycling, and suppression of tumorigenesis. To exert those pleiotropic effects, VDR, after dimerization with RXR (retinoic acid receptor X) and translocation to the cell nucleus, interacts with the D receptor-interacting protein (DRIP), the steroid receptor co-activator (SRC) family proteins (with SRC2 and 3 expressed in keratinocytes), β-catenin, and the inhibitor hairless protein (Hr) (Bikle 2011b).

The gradient of calcium level in the epidermis, defined by its low content in the basal layer and the highest level in the corneal layer, determines the expression of several genes required for differentiation of the epidermis and proper epidermal barrier formation. The calcium-stimulated keratinocyte differentiation requires the activity of several proteins including calcium receptor (CaR), phospholipase PLC-γ 1, and SRC kinases (Bikle 2011a). Calcitriol, the active form of vitamin D_3, was shown to stimulate the expression of CaR, which subsequently sensitizes keratinocytes to calcium. Moreover, calcitriol is involved in the regulation of synthesis, processing, and trafficking of glycosylceramides, which are important for the regulation of epidermal permeability barrier. In human keratinocytes calcitriol induces the expression of loricrin, filagrin, and phospholipase PCL-γ1 and, synergistically with Ca^{+2}, increases the expression of involucrin and transglutaminase K (enzyme required for the cornification of keratinocytes). Antiproliferative effects of calcitriol are also mediated by inhibiting c-myc, cyclin D1, p21, and p27 (Bikle 2011b).

The pleiotropic effects of vitamin D_3 on skin physiology cannot be solely explained by the differential expression of VDR, which is the highest in the basal layer of the epidermis. As previously mentioned, the activity of VDR is modulated by its interactions with DRIP coactivator complex and SRC proteins. Moreover, it was also shown that in undifferentiated keratinocytes VDR preferentially binds to DRIP coactivator complex in contrast to the more differentiated epidermal cell layers, where VDR interacts with SRC proteins (Bikle 2010).

Vitamin D is also involved in the regulation of skin immune responses by stimulating the expression of IκBα, a NF-κB inhibitor, which leads to the retention of NF-κB in the cytoplasm and its subsequent degradation (Janjetovic et al. 2009, 2010; Lu et al. 2004; Reichrath 2007). Thus, vitamin D and its derivatives inactivate the major transcription factor responsible for the transcription and release of many inflammatory mediators. On the other hand, vitamin D can stimulate innate immune responses, including production of antimicrobial peptides (Gombart et al. 2005; Reichrath 2007).

Basal and squamous cell carcinomas are the most frequent types of human cancer. Because vitamin D and its derivatives inhibit proliferation and stimulate differentiation of keratinocytes, its possible use in the treatment of skin cancer and hyperproliferative diseases of the skin, including psoriasis, has been suggested (reviewed by Bikle 2011a; c; Holick 2008; Lehmann et al. 2004). It is likely that hedgehog or β-catenin pathways can serve as targets of the anticarcinogenic activity of vitamin D in the skin (Bikle 2011c; Tang et al. 2011; Teichert et al. 2011). Furthermore, calcitriol and other derivatives of vitamin D show anti-melanotic activities, and defective expression of VDR correlates with poor prognosis of melanoma (reviewed by Berwick et al. 2005; Brozyna et al. 2011; Field and Newton-Bishop 2011; Janjetovic et al. 2011; Pinczewski and Slominski 2010; Slominski et al. 2011b).

8.4 Conclusions

We discovered a novel pathway initiated by the enzymatic action of CYP11A1 on 7DHC (provitamin D3), vitamin D3 (Slominski et al. 2004d, 2009c), and plant-derived, i.e., provitamin D2 (Slominski et al. 2005g, 2006d; Tuckey et al. 2011). The CYP11A1 metabolites of 7DHC, and hydroxyproducts of 7DHP generated by enzymes of classical steroidogenic pathway to produce 5,7-dienal steroids, can be converted by ultraviolet B radiation (UVB) to novel vitamin D hydroxyderivatives, androsta-calciferols, and pregna-calciferols, which are biologically active in keratinocytes and melanoma cells (Janjetovic et al. 2010; Slominski et al. 2005f, 2010, 2011b; Zbytek et al. 2008; Zmijewski et al. 2011). In addition, lumisterol- and tachysterol-like compounds are produced in the skin by the action of UVB. The biological activity of the vitamin D-derived compounds is defined by their chemical structure and cell lineage. Since P450scc is expressed in the skin, we propose that novel secosteroidal pathways may affect skin biology with potent therapeutical implications. Moreover, body's global homeostasis can potentially be affected by skin-derived secosteroids as well as by CYP11A1-mediated hydroxylation of hydroxyvitamin D in organs or tissues which exhibit high activity of this enzyme.

The associations of vitamin D levels with various pathological states, including osteoporosis, rickets, coronary heart disease, and carcinogenesis, have recently become the subject of intensive clinical investigations. It is hoped that the results of these studies will provide sufficient data to define the requirements for vitamin D dietary supplementation. Also, the newly described low calcemic vitamin D derivatives, such as 20-OH D_3, may serve in the future for anticancer therapy.

Chapter 9
Equivalent of Hypothalamic–Pituitary–Thyroid Axis

9.1 Overview

The hypothalamic–pituitary–thyroid (HPT) axis represents the central regulatory mechanism of cellular metabolism including protein, carbohydrate, and lipid catabolism. The HPT axis is also involved in cell differentiation and proliferation as well as morphogenesis. Tight control of thyroid hormone activity is orchestrated by hypothalamic synthesis of the thyroid-releasing hormone (TRH), which activates its receptors (TRH-R1) in the anterior pituitary, followed by thyroid-stimulating hormone (TSH) secretion. TSH stimulates the production of 3,5,3′-triiodothyronine (T3) and thyroxine (T4) by thyreocytes. Elevated serum levels of thyroid hormones inhibit TRH and TSH synthesis by hypothalamus and pituitary, respectively (Fig. 9.1). Although T4 is the major thyroid hormone found in circulation, T3 is a fully active hormone since deiodinases (type 2 and 3) are expressed in a variety of tissues, including skin, and facilitate the conversion of T4 to T3 (Fig. 9.1). Skin is a nonclassical target for TSH, TRH, and thyroid hormones.

9.2 HPT in the Skin

The expression of the molecular elements of the HPT axis (genes for TSH-R1, TSH, TRH, TRH-R1, deiodinases 2 and 3, thyroglobulin, and sodium iodide transporter) in human skin and functional TSH receptors in keratinocytes and malignant melanocytes was first shown by Slominski et al. (2002e). Follow-up studies demonstrated TRH and TSH receptors' expression in various skin cell types including keratinocytes, melanocytes, fibroblasts, and hair follicles (Bodo et al. 2010; Cianfarani et al. 2010; Gaspar et al. 2010; Pattwell et al. 2010; van Beek et al. 2008), which gave ground to the concept of a cutaneous HPT axis that would show similarities to and differences with the central HPT axis (Slominski et al. 2002e). The presence of TSH and TRH receptors explains the phenotypic changes in

A.T. Slominski et al., *Sensing the Environment: Regulation of Local and Global Homeostasis by the Skin's Neuroendocrine System*, Advances in Anatomy, Embryology and Cell Biology 212, DOI 10.1007/978-3-642-19683-6_9, © Springer-Verlag Berlin Heidelberg 2012

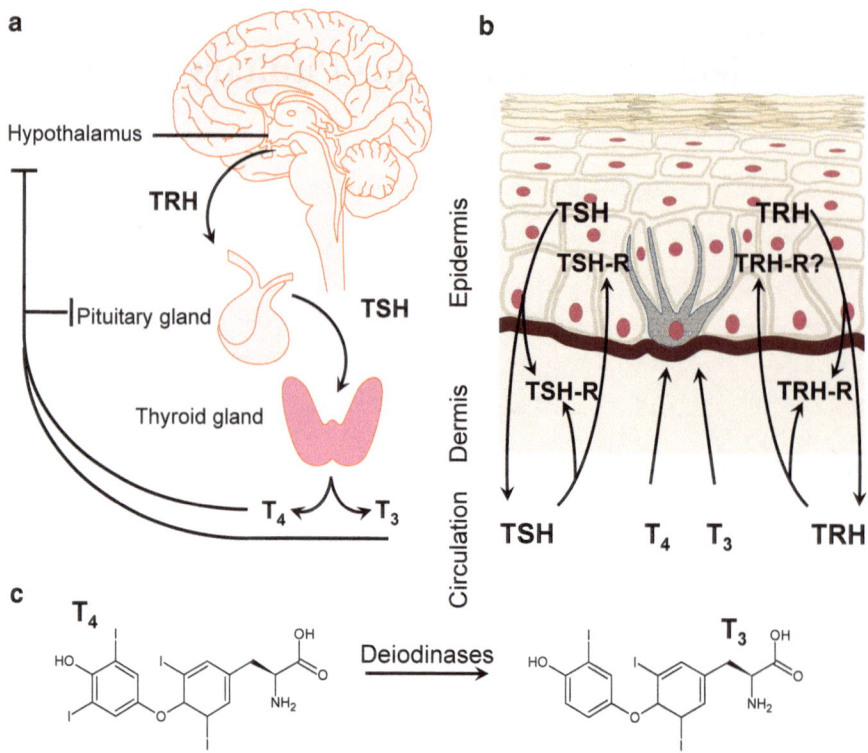

Fig. 9.1 Structure and functions of an equivalent of the hypothalamic–pituitary–thyroid axis (HPT) in the skin

epidermal and dermal cells treated with TRH and TSH (Bodo et al. 2010; Cianfarani et al. 2010; Gaspar et al. 2010; Pattwell et al. 2010; Slominski et al. 2002e; van Beek et al. 2008). The interaction of T3 with its receptors (TRα and TRβ) affects epidermal differentiation and enhances its responsiveness to growth factors (Billoni et al. 2000; Slominski and Wortsman 2000). These effects of T3 are particularly important for the function of sebaceous, eccrine, and apocrine glands, growth of hair follicles, and synthesis of proteo- and glycosaminoglycans by dermal fibroblasts. For instance, both types of T3 receptors can regulate keratinocytes' proliferation, differentiation, and immune activity (Contreras-Jurado et al. 2011). Results of the latter study suggest that thyroid hormones acting via their receptors can inhibit skin inflammation, most likely by inactivating specific transcription factors: AP-1, NF-κB, and STAT3 (Contreras-Jurado et al. 2011). T4 stimulates the proliferation of hair follicle keratinocytes and T3 inhibits their apoptosis (van Beek et al. 2008). Moreover, thyroid hormone receptors might suppress invasiveness and metastatic ability of skin tumors as shown in mouse knockout models (Martinez-Iglesias et al. 2009). Thyroid hormones may also affect hair follicle stem cells, since T3 and T4 were found to induce differentiation and apoptosis, and inhibit clonal growth of hair follicle epithelial stem cells (Tiede et al. 2010). The activity of deiodinases enables cutaneous conversion of T4 into T3, what plays a role in the regulation of the proliferation of keratinocytes and dermal fibroblasts in vitro and in vivo (Huang et al. 2011; Safer et al. 2009).

The effects of thyroid hormones in the skin are well pronounced in case of thyroid gland pathology including autoimmune thyroid disease (Cianfarani et al. 2010; Slominski and Wortsman 2000). Dermal manifestations of hyperthyroidism include erythema, palmoplantar hyperhidrosis, acropathy, and infiltrative dermopathy. Moreover, Graves' disease also may be associated with generalized itching, chronic urticaria, presence of alopecia areata, and vitiligo (Doshi et al. 2008; Ingordo et al. 2011; Kasumagic-Halilovic et al. 2011). Skin manifestations of thyroid disorders are in part correlated with elevated serum levels of thyroglobulin (Tg), thyroperoxidase (TPO), and thyroid-stimulating hormone receptor (TSH-R) antibodies (Cianfarani et al. 2010; Slominski and Wortsman 2000). Conversely, in hypothyroidism the skin is cool and dry with pasty appearance; hair are commonly dry, coarse, and brittle with up to a 50% probability of diffuse or partial alopecia development (reviewed by Slominski and Wortsman 2000). Several studies also underlined the significance of TSH and TRH in the human skin and their influence on hair physiology (Bodo et al. 2010; Cianfarani et al. 2010; Gaspar et al. 2010). The physiological activity of TRH was demonstrated in human hair follicles' organ cultures expressing active TRH-R1 receptor. TRH stimulated hair shaft formation, prevented apoptosis, increased proliferation of hair matrix keratinocytes, and prolonged the anagen phase of hair growth cycle (Gaspar et al. 2010; van Beek et al. 2008). TRH can also stimulate hair follicle pigmentation (Gaspar et al. 2011), probably by direct activation of melanocortin type 1 receptor (MC1-R), confirming our previous hypothesis (Slominski et al. 2002e, 2005b). It was shown that TSH acting via TSH-R1 receptor increased cAMP production by human keratinocytes and human and hamster melanoma cells (Slominski et al. 2002e) and enhanced the proliferation of epidermal keratinocytes and dermal fibroblasts (Bodo et al. 2010). Furthermore, TSH stimulated cyclic AMP production and altered the expression of several genes in human hair follicles and dermal papilla fibroblast cultures from normal female skin, however, did not affect hair growth and pigmentation (Bodo et al. 2009; van Beek et al. 2008). Expression of the TSH-R protein was detected in a wide panel of melanocytic lesions including melanoma. TSH activated the mitogen-activated protein kinase (MAPK) pathway and stimulated proliferation of melanoma cells, however, not melanocytes (Ellerhorst et al. 2010). Furthermore, TRH at low concentrations stimulated melanoma growth but not melanocytes' proliferation and its expression was increased in dysplastic nevi in contrast to benign nevi and was expressed in 63% of melanoma samples (Ellerhorst et al. 2006). Interestingly, other authors reported that suppression of MAP kinase and PI3K/Akt pathways, while leading to inhibition of cell proliferation, induced thyroid genes' expression including TSH-R and sodium/iodide symporter, which led to increased iodine uptake by melanoma cells (Hou et al. 2009).

Recent studies demonstrated the expression of thyroid factor-1, thyroglobulin (Bodo et al. 2009), and thyroperoxidase in the human skin (Cianfarani et al. 2010). T3 and T4 were also shown to modulate the expression of cytokeratins 6 and 14 genes and downregulate TGF-β2 expression in hair follicles (van Beek et al. 2008). Thyroid hormones also stimulated hair pigmentation (van Beek et al. 2008) and mitochondrial function by upregulating the mRNA level of mitochondria-selective cytochrome-c-oxidase subunit 1 (MTCO1) and significantly increasing complex I and IV (cytochrome-c-oxidase) activities in the epidermis (Poeggeler et al. 2010).

It is well established that thyroid dysfunction alters skin physiology, but expression of the equivalent of HPT axis in the skin raises also the possibility of a cross-talk between local and systemic counterparts as it was demonstrated in amphibians (Vaudry et al. 1999). These interactions might have long-range consequences, especially for the regulation of global homeostasis, evolution of skin stress response systems, and development of thyroid-related autoimmune diseases (reviewed by Slominski et al. 2008b). From the clinical point of view, pathological exposure of TSH-R to immune cells in keratinocytes damaged by UVR or fibroblasts damaged during inflammation (Slominski et al. 2002e) can induce either production of anti-TSH-R antibodies leading to the uncontrolled stimulation of the thyroid gland or generation of anti-TSH-R clones of T lymphocytes—leading to immune destruction of the thyroid. These concepts, originally proposed by Slominski et al. (2002e), have been recently reemphasized defining a role of skin in thyroid autoimmune diseases (Cianfarani et al. 2010).

9.3 Conclusions

Different elements of the HPT axis are expressed in the skin and this expression is skin cell-type dependent. The expression of individual or networked HPT elements can regulate skin phenotype in a differentiated and context-dependent manner (Fig. 9.1). Possibly, communication between the cutaneous HPT axis and other local neuroendocrine networks as well as with central coordinating centers takes place and affects global homeostasis (Figs. 1.1 and 9.1).

Chapter 10
Cutaneous Opioid System

10.1 Overview

Endogenous opioid peptides derive from four different precursor proteins. Proopiomelanocortin (POMC) yields ACTH and endorphins, mainly β-endorphin (β-END). Proenkephalin (PENK) generates enkephalins (ENK), predominantly Met-enkephalin (MENK) and Leu-enkephalin (LENK). The proteolysis of prodynorphin (PDYN) results in the formation of dynorphin A (DYN A) and B (DYN B) (Przewlocki 2004; Przewlocki and Przewlocka 2005). Other POMC derivatives are endomorphins which are the cleavage products of a larger precursor molecule that yet has not been identified (Fichna et al. 2007).

Three classes of the Gi/Go/Gq-coupled opioid metabotropic receptor (OR) family have been identified: mu (μ, MOR), delta (δ, DOR), and kappa (κ, KOR). In addition, an orphan opioid-like nociceptin receptor (NOP), which has a 70% sequence homology with other opioid receptors, was also characterized (Jordan and Devi 1998; Jordan et al. 2000; Przewlocki and Przewlocka 2005; Salemi et al. 2005). Activation of ORs inhibits cAMP signaling as well as alters voltage-gated Ca^{2+} channel function and activates K^+ inwardly rectifying channels. The endomorphins bind to MOR with the highest affinity, ENK preferentially binds to the DOR, and DYN favorably binds with KOR (Przewlocki 2004). The ligand-receptor affinity presents in the following order: β-END μ, δ > κ; LENK/MENK δ, μ; DYNA κ > μ > δ (Jordan et al. 2000; Tominaga et al. 2007).

10.2 Opioid Peptides in the Skin

10.2.1 β-Endorphin

It was shown that human and animal skin and/or cultured skin cells such as keratinocytes and melanocytes (normal and pathological) have the capability to transcribe and translate the precursor opioid protein genes (Nissen and Kragballe

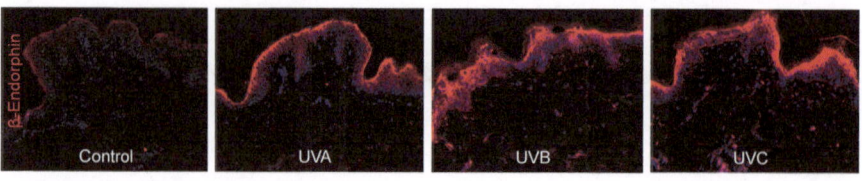

Fig. 10.1 Wavelength-dependent UV stimulation of β-END expression in the epidermal layer of human skin. CY™3 positive (*red*) signals correspond to β-END immunoreactivity (methods described in Skobowiat et al. 2011)

1997; Polakiewicz et al. 1992; Salemi et al. 2005; Sikand et al. 2011; Skobowiat et al. 2011; Slominski et al. 1992, 1993b, 2000c, 2011c; Zagon et al. 1996). The POMC gene and protein and POMC-derived peptides (ACTH, α-MSH, and β-END) were detected in epidermis, dermis, and adnexa and, additionally, can be released from the cutaneous nerve endings (Hasse et al. 2007; Mazurkiewicz et al. 2000; Slominski et al. 1992, 1993b, 1998a, 2000c; Tobin and Kauser 2005a; Wintzen and Gilchrest 1996). The cutaneous expression of POMC was first found in rodent (hamster and mouse) melanomas (Slominski 1991) and in mouse (Slominski et al. 1992) and human (Slominski et al. 1993c) skin, where the β-END antigen was detected as well. β-endorphin stimulates keratinocyte migration in vitro (Tominaga et al. 2007), induces epidermal and follicular melanogenesis (Kauser et al. 2004), and also controls hair growth, wound healing, and cellular differentiation (Schmelz and Paus 2007). Furthermore, β-END increases and modulates the number of dendritic processes of hair follicle melanocytes (Kauser et al. 2004). Some authors reported that plasma β-END levels increase after UV exposure, which would explain the euphoric behavior observed in beachgoers (Fallahzadeh and Namazi 2009). Other authors could not observe this correlation (Wintzen et al. 2001). Recently, UVR-induced increase of β-END level has been observed in ex vivo maintained human skin (Skobowiat et al. 2011) (Fig. 10.1).

10.2.2 Proenkephalin

The expression of the *PENK* gene and the subsequent processing of PENK to MENK/LENK (in a cell type-dependent manner) were demonstrated in the skin by RT-PCR, Western blotting, immunocytochemistry, time-of-flight/liquid chromatography, and mass spectroscopy (Slominski et al. 2011c). PENK immunoreactivity was markedly restricted to differentiating keratinocytes of the stratum spinosum and granulosum, whereas proliferating basal keratinocytes did not exhibit this immunoreactivity (Slominski et al. 2011c; Zagon et al. 1996). Furthermore, physical (UVB) and biological (lipopolysaccharide) stressors demonstrated cell type-specific time- and dose-dependent stimulation of *PENK* gene expression (Slominski et al. 2011c). Also, fetal mesenchymal skin cells expressed and produced significant amount of PENK, indicating its association with cell proliferation

(Polakiewicz et al. 1992). MENK released from rat Merkel cells acted in an autocrine/paracrine fashion by the inhibition of cell granules' release via decrease of intracellular Ca^{2+} concentration (Tachibana and Nawa 2005). ENK plays a role in the differentiation of epidermal keratinocytes and has direct antimicrobial activities which contribute to the skin protective barrier against noxious factors (Nissen and Kragballe 1997; Slominski et al. 2011c).

10.2.3   Dynorphins

PDYN and DYN A are expressed in human skin cells as well as in cutaneous nerve fibers (Grando et al. 1995; Hassan et al. 1992; Salemi et al. 2005; Taneda et al. 2011; Tominaga et al. 2007). They are responsible for pilomotor activity (Gibbins 1992) and nociception. Enhanced production of DYN A was found in atopic dermatitis (Tominaga et al. 2007). Endomorphin 1 and 2 immunoreactivity was found in nerve fibers of the rat skin; however, the physiological function of these peptides still has to be determined (Barr and Zadina 1999; Borzsei et al. 2008).

10.3   Opioid Receptors and Their Function in the Skin

Previous studies showed that not only opioid peptides but also their receptors were expressed in the skin (Grando et al. 1995; Nissen and Kragballe 1997; Salemi et al. 2005; Tachibana and Nawa 2005). These findings define skin as an active environment for opioid action. Ectoderm-derived cells, e.g., keratinocytes and melanocytes, express ORs, however, at much lower level (by a factor of 200–20,000) than neurons. MOR showed a stronger expression than DOR in keratinocytes, but an opposite expression pattern was found in mesenchyme-derived fibroblasts (Bigliardi et al. 1998). Furthermore, KOR was detected in fibroblasts and mononuclear blood cells of normal human skin and DOR was expressed in fibroblasts isolated from human skin (Salemi et al. 2005). Enhanced expression of DOR and KOR in the skin justifies the exploration of novel δ and κ acting compounds as specific targets for future opioid therapy. OR-induced signaling can affect cell differentiation, migration, as well as cytokeratin and cytokine expression in human epidermis. Thus, opioid receptors may be involved not only in the regulation of normal skin homeostasis but also in wound healing and scar formation (Bigliardi et al. 2009).

Opioids are best known for their antinociception, which can also be initiated by the activation of ORs outside the central nervous system (Salemi et al. 2005). Their inhibitory activities are related to the ligand activation of Gi,o,q proteins which downregulates adenyl cyclase activity and, eventually, inhibits protein phosphorylation (Chizhmakov et al. 2005). In addition, modifications of ion channels' activity have further expanded the spectrum of opioid-induced biological action. All ORs couple to various Ca^{2+} channels and are known to inhibit their activity. Stimulation of ORs also increases potassium conductance across the cellular membranes (Jordan and Devi 1998; Jordan et al. 2000). Opiates increase the release of dopamine via presynaptic inhibition of GABA release, which leads to a reward response, an effect observed in tanning addicts (Harrington et al. 2006; Nolan and Feldman 2009; Przewlocki 2004). Inhibitory action of opiates on noradrenergic activity is mediated by MOR, and its activation enhances CRF production and release (Armario 2010). Opioids exert immunomodulatory effects in peripheral tissues including skin by stimulating lymphocyte proliferation, antibody production, T and NK cell activity, and chemotaxis of macrophages and granulocytes (Fallahzadeh and Namazi 2009; Jankowska and Schomburg 1998; Jordan and Devi 1998; Mousa et al. 2007).

Opioids produced by skin cells may probably act as para- and autocrine modulators that influence gene expression. Additionally, they may enter the systemic circulation being transported by dermal veins, and interact with specific receptors localized on cutaneous nerve fibers (Bigliardi et al. 2009; Borzsei et al. 2008; Slominski and Wortsman 2000; Slominski et al. 2000c). The last property is required for their analgesic and anti-inflammatory activity. Transduced nervous signals from peripheral tissues could be conveyed to dorsal root ganglia (DRG) sensory neurons. Thus, via synaptic inhibition neuropeptides like substance P (SP) and calcitonin gene-related peptide (CGRP), responsible for pain and inflammation, would be downregulated and not delivered to skin (Borzsei et al. 2008; Slominski and Wortsman 2000; Tobin 2006). Furthermore, the ascending activator signals could be transferred via the nucleus of the solitary tract (NTS) to thalamus, hypothalamus, especially to paraventricular and arcuate nuclei, and amygdala where they could exert their systemic actions (Slominski et al. 2008b) and enhance the activity of the reward system (Armario 2010; Przewlocki 2004; Przewlocki and Przewlocka 2005; Slominski 2005; Slominski and Wortsman 2000) (Fig. 10.2).

10.4   Opioid System and Pruritus

It has been known for decades that analgesia resulting from MOR activation induces itch, whereas MOR antagonists, such as naltrexone, inhibit itch (Yosipovitch 2010). It is widely accepted that the activation of KOR signaling suppresses, while that of MOR stimulates, itch (Bigliardi et al. 2009; Roosterman et al. 2006). This has led to the opioidergic system being targeted by new antipruritic medications (Patel and Yosipovitch 2010; Schmelz 2010). A novel KOR agonist, nalfurafine/TRK-820, revealed antipruritic activity in morphine-, histamine-, and substance P-induced animal scratching models (Ko and Husbands 2009). The number of nerve fibers entering the epidermis tended to increase in approximately 40% of psoriatic patients claiming itch sensation compared to healthy controls (Taneda et al. 2011). There were no differences in epidermal number of MOR and β-END amount; however, levels of KOR and DYN A were

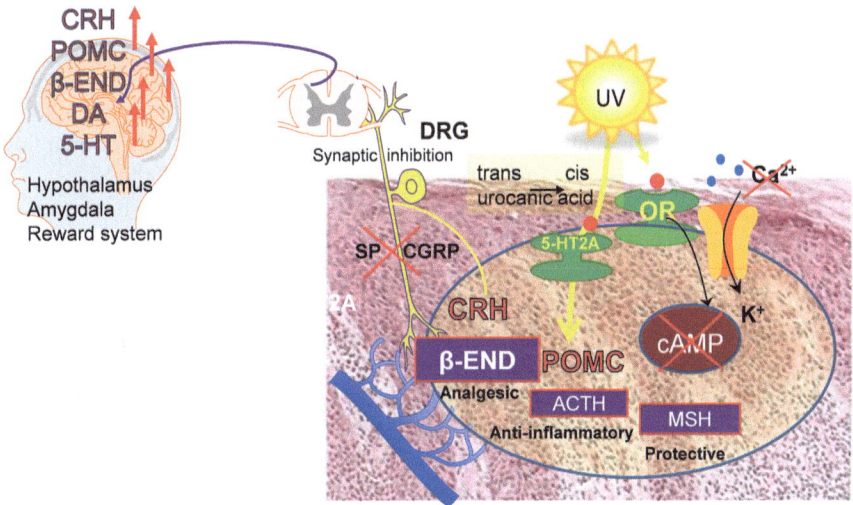

Fig. 10.2 Hypothetical pathway of addictive activities of the cutaneous neuroendocrine system induced by UV stimulation

significantly decreased between healthy controls and psoriatic patients (Taneda et al. 2011). Interestingly, opioids were shown to act upon capsaicin-sensitive nerve fibers and inhibit the release of inflammatory neuropeptides such as SP, neurokinin A, and CGRP, i.e., neuropeptides that are indirectly involved in eliciting pruritus by releasing the pruritogen—histamine (Roosterman et al. 2006; Stander et al. 2002).

It was shown that UV-induced keratinocyte-delivered nerve growth factor (NGF), upon retrograde transport from skin toward dorsal root ganglion, enhanced the expression of neuropeptides SP and CGRP and upregulated the number of MORs localized on cutaneous sensory nerve fibers (Mousa et al. 2007; Roosterman et al. 2006). Recent studies have provided evidence that, indeed, there are itch-specific receptors in the skin. Using in vitro binding assays, it was observed that the proteolytically cleaved product of proenkephalin A, the bovine adrenal medulla peptide 8–22 (BAM8–22), potently activated the Mas-related G-protein-coupled receptors (Mrgprs) in an opioid-independent mechanism (Sikand et al. 2011).

10.5 Opioids and Addiction

UV exposure during indoor tanning damages DNA, thereby leading to premature skin aging and the development of skin cancer as well as malignant melanoma (Harrington et al. 2006). The majority of beachgoers reported behaviors consistent with those of an addictive disorder like continuation of tanning despite attempts to stop, persistent tanning in the presence of adverse reactions, and the neglect of other responsibilities in order to maintain a tan (Keen et al. 2008). Psychological dependence is suggested by tanners' reports of relaxation and positive mood effects as a result of UV exposure. These observations may be explained by UV-induced production of cutaneous β-END (Skobowiat et al. 2011) resulting from local transcription, translation, and further cleavage of POMC leading to the production of β-END (reviewed by Slominski 2003; Slominski et al. 1993b, 2000c).

Solar UV energy adsorbed by the epidermis also results in the transformation of a chromophore-like trans-urocanic acid to its cis-isomer, which reportedly could have an agonistic activity on serotonin receptor 2A (5HT2A) (Walterscheid et al. 2006). Thus, UV light can also alter cutaneous serotoninergic system with subsequent effects on the CNS affecting the mood (Harrington et al. 2011; Kourosh et al. 2010; Slominski et al. 2005c). Upon UV exposure, 5-HT could be transported from the skin via its blood vessels or activate ascending nerve fibers which affect brain activity (Nordlind et al. 2008; Slominski et al. 2005c). Specifically, brain striatal regions, including the nuclei accumbens, caudate and putamen, could be activated during UVR exposure (reviewed by Kourosh et al. 2010). Frequent tanning may involve CNS reward and/or reinforcement over the often-stated goal of "getting a tan." The ventral striatum (or nucleus accumbens) activation is typically associated with drug-induced reward observed in cocaine and nicotine smokers (Harrington et al. 2006). In fact, the term "psychodermatology" is being used to describe the mind–skin connection (Reich et al. 2010). From an evolutionary perspective, it is reasonable to suggest that sunlight may have central rewarding properties (Figs. 1.1 and 10.2) given the importance for human health of UV-mediated vitamin D synthesis.

10.6   Conclusions

Opioids constitute a heterogeneous family of active peptides which play important roles in cutaneous nociception, immunomodulation, signal transduction, and evoking or attenuating of pruritus, depending on differential receptor activation. Their main inhibitory properties are related to the inhibition of cell membrane calcium channels that suppress the release of proinflammatory and pain transmitters like SP and CGRP. Both exogenously applied and endogenous opioids interact with a whole range of receptors and contribute to the neuroendocrinological functions of the skin, also at the systemic level (Figs. 1.1, 1.2 and 10.2).

Chapter 11
Cutaneous Endocannabinoid System

11.1 Overview

Endocannabinoids (ECS) constitute lipid mediators (amides, esters, and ethers of long chain polyunsaturated fatty acids) which act similarly to the exogenous Δ^9 tetrahydrocannabinol (THC; the main psychoactive ingredient of the plant *Cannabis sativa*) and are produced in humans and animals (Maccarrone et al. 2003; Rahn and Hohmann 2009). The cutaneous ECS system is fully functional due to the expression of ECS and their receptors as well as ECS-degrading enzymes. All the components of the skin ECS system were shown to modulate the proliferation, differentiation, growth, and apoptosis of various skin cell types as well as tumorigenesis and local cytokine production (reviewed by Biro et al. 2009; Kupczyk et al. 2009; Toth et al. 2011). ECS are synthesized "on demand" by receptor-stimulated cleavage of membrane lipid precursors and are not stored in synaptic vesicles which distinguishes them from typical neurotransmitters. ECS reuptake may be facilitated by a transporter that has not been cloned yet; however, pharmacological inhibitors of ECS transport have nonetheless been developed (Guindon and Hohmann 2009). The lipophilic nature of ECS allows them to activate various enzymes in cytosol and membraneous compartments (Kupczyk et al. 2009).

The most extensively studied ECS are *N*-arachidonoylethanolamide (anandamide/AEA), 2-arachidonoylglycerol (2-AG), and *N*-palmitoylethanolamide (PEA), while their main catabolic enzymes are fatty acid amid hydrolase (FAAH) and the monoacylglycerol lipase (MAGL).

A.T. Slominski et al., *Sensing the Environment: Regulation of Local and Global Homeostasis by the Skin's Neuroendocrine System*, Advances in Anatomy, Embryology and Cell Biology 212, DOI 10.1007/978-3-642-19683-6_11, © Springer-Verlag Berlin Heidelberg 2012

11.2   Endocannabinoid Receptors and Mechanisms of ECS Action

Two receptor types for cannabinoids, CB1 and CB2, have been identified beyond doubt; however, some researchers assume the existence of a third one, CB3, which has not been cloned yet (Kupczyk et al. 2009). The amino acid sequences of the CB receptors are conserved in many species, including humans (Kupczyk et al. 2009). CB1 is predominantly expressed in the central nervous system and other tissues, while CB2 was mainly found in non-neuronal cells and tissues related to the immune system, like lymphocytes, macrophages, spleen, and thymus. In the skin, both receptor types are expressed on nerve endings, mast cells, keratinocytes, and adnexa (Biro et al. 2009; Maccarrone et al. 2003). Classical activation of CBs, which belong to the coupled Gi/o family of G proteins, inhibits N- and P/Q-type Ca^{2+} channels, activates A-type, inwardly rectifying potassium conductance channels, and inhibits M-type potassium channels. Furthermore, binding of CB2 receptor ligands modulates MAPK activity resulting ultimately in decreased cAMP production and suppression of neuronal excitability and transmitter release (reviewed by Irving et al. 2002). Additionally, CB1 activation can inhibit conductance of serotonin receptor 3 (5-HT3) ion channels, modulate the production of NO, alter voltage-sensitive sodium channel activity, and activate the Na^+/H^+ exchanger (Guindon and Hohmann 2009; Rahn and Hohmann 2009). It was shown that CB1 and transient receptor potential vanilloid-1 (TRPV-1) are co-localized on sensory nerve endings in the skin (Akerman et al. 2004) which suggests that ECS can also activate TRPV-1 receptors in the skin (Karst et al. 2010; Maccarrone et al. 2003). Anandamide acts at vanilloid receptors and blocks directly the background K^+ channel affecting NMDA transmission in the brain (reviewed by Irving et al. 2002). CB1 receptors are localized presynaptically on GABA-ergic and glutamatergic interneurons (Kupczyk et al. 2009). CB1 activation results in a decreased release of neurotransmitters such as GABA (γ-aminobutyric acid) and glutamate. This retrograde signaling mechanism suggests an important modulatory role of ECS in controlling neuronal excitability and homeostasis. ECS mediate habituation to stress by restraining HPA axis response and maintaining body's homeostasis (Finn 2010).

11.3   Effects of ECS on Proliferation and Differentiation of Keratinocytes

CB1 agonists inhibited proliferation of cultured human epidermal keratinocytes and anandamide (AEA) markedly suppressed cell growth and induced dose- and CB1-dependent apoptosis in human HaCaT keratinocytes (Paradisi et al. 2008; Toth et al. 2011). AEA inhibited hair shaft elongation and proliferation of hair matrix keratinocytes (Telek et al. 2007). Cannabinoids also induced intraepithelial apoptosis and premature hair follicle regression (characteristic signs of catagen transformation in hair follicles)—processes that were inhibited by a selective CB1 antagonist (Pucci et al. 2011; Toth et al. 2011). CB1 is expressed in a hair cycle-dependent manner and negatively regulates human hair growth in an autocrine--paracrine manner. Indeed, it was shown that CB1 antagonists induced hair growth in mice (Srivastava et al. 2009).

Differentiating human keratinocytes had decreased levels of endogenous AEA due to increased degradation of this lipid (Maccarrone et al. 2003). Moreover, it was shown that exogenous AEA inhibited keratinocyte differentiation in vitro acting via a CB1-dependent mechanism that involved inactivation of protein kinase C, activating protein-1 (AP-1), and transglutaminase (Maccarrone et al. 2003). High expression of CB1 in epidermal granular and spinous layers suggests participation of ECS in keratinocytes' differentiation (Stander et al. 2005).

11.4 Effects on Tumorigenesis

Various human skin tumors (e.g., basal and/or squamous cell carcinoma) express both CB1 and CB2 receptors (Casanova et al. 2003; Zheng et al. 2008). Local administration of synthetic CB1 and CB2 agonists inhibited growth of skin malignant tumors in nude mice by increasing intra-tumor apoptosis and impairing tumor vascularization (Casanova et al. 2003). ECS also inhibited in vivo growth of mouse melanomas that expressed CB1 and CB2 by decreasing growth, proliferation, angiogenesis, and metastasis formation, while increasing apoptosis (Blazquez et al. 2006). It was also found that human squamous cell carcinoma overexpressed CB2 at the both mRNA and protein levels (Zhao et al. 2010). UVB exposure of experimentally induced papilloma in mouse skin led to a local activation of CB1/2 receptors. While the absence of the functional CB1/2 receptors in double knock-out mice resulted in a resistance to UVB-induced inflammation and a marked decrease in UVB-induced skin carcinogenesis. Thus, the CB1/2 receptors play a key role in UV-induced inflammation and skin cancer development (Zheng et al. 2008).

11.5 Effects on Inflammation

CB2 receptor activation, in general, mediates immunosuppressive effects, which limit inflammation and associated tissue injury in large number of pathological conditions. Interestingly, stimulation of CB2 receptors in immune cells after initial decrease in cAMP production may lead to a sustained, pronounced increase in cAMP levels, which results in the suppression of T-cell receptor signaling through a cAMP/PKA/Csk/Lck pathway (reviewed by Pacher and Mechoulam 2011). Recent studies have also revealed that ECS may affect proliferation and apoptosis of T and

B lymphocytes, inflammatory cytokine production, and immune cell activation by inflammatory stimuli (e.g., LPS), macrophage-mediated killing of sensitized cells, chemotaxis, and inflammatory cell migration (Sanchez and Garcia-Merino 2012). In a mouse model of contact allergy, the cutaneous inflammation has been suppressed by local administration of Δ^9 tetrahydrocannabinol (THC) and CB agonists (Karsak et al. 2007). In murine dermatitis elevated 2-AG (2-arachidonoylglycerol) levels were observed, and inflammatory symptoms were markedly attenuated by CB2 (but not CB1) agonists (Oka et al. 2006). Decreased dermal fibrosis (bleomycin-induced) and inflammation were observed upon treatment with a CB2 agonist, suggesting a potential therapeutic application of selective CB2 agonists in early inflammatory stages at the local and systemic levels (Akhmetshina et al. 2009).

11.6 Effects on Pain Sensation

Antinociceptive activity of cannabinoids has been one of the main reasons of their worldwide usage in many communities. Recently described mechanisms of antinociceptive action of ECS involve the specific CB's activation as well as interaction with other receptors and pathways related to pain sensation at the central (Connell et al. 2006; Irving et al. 2002) and peripheral (Finn 2010; Karst et al. 2010; Amaya et al. 2006) levels. Activation of CBs changes cellular Ca^{2+} and K^+ conductance and decreases cAMP levels leading to an inhibition of neurotransmitter's release which is involved in pain sensation (Rahn and Hohmann 2009; Akerman et al. 2004).

Antinociceptive effects of CB1/2 agonists, and FAAH inhibitors, which prolong the action of ECS, have been reported. It was shown in rats that CB2 agonists stimulate release of β-END from keratinocytes, at the local mu-opioid related fashion, to inhibit nociception in the skin (Ibrahim et al. 2005). The application of electroacupuncture and CB2 agonist reduced inflammatory pain due to increased POMC mRNA and β-END levels (via mu-opioid receptor mechanism) in inflamed rat skin (Su et al. 2011). Increased mobilization and activity of ECS in amygdala, observed after foot shock stress in rats, lead to an attenuation of nociceptive pathways due to CB1 activation (Connell et al. 2006). Furthermore, cannabis (inhaled as aerosol) augmented opioid analgesia through a pharmacodynamic mechanism, in experimental study performed in humans (Abrams et al. 2011). Elevated levels of ECS in models of chronic pain are likely to counteract increased neuronal activity driven by afferent nociceptive inputs. ECS-induced inhibition in neurotransmission may modulate central sensitization observed during pain stimuli (Connell et al. 2006; Abrams et al. 2011, reviewed by Karst et al. 2010). The cutaneous application of synthetic CB agonist was shown to reverse inflammatory thermal hyperalgesia evoked by Freund's adjuvant application in rats (Amaya et al. 2006). Elevated ECS level, evoked by blocking of catabolic (mainly FAAH) enzymes activity, is effective and desired feature in decreasing pain perception during osteoarthritis in rats (Sagar et al. 2010). Extended ECS action, due to FAAH deactivation, is likely to be more beneficial compared to direct activation of CB1 receptors (reviewed by Karst et al. 2010; Rahn and Hohmann 2009). Interestingly, both AEA and 2-AG can be metabolized by cyclooxygenase-2, which may contribute to the pain relieving properties of nonsteroidal anti-inflammatory drugs that act by inhibiting the cyclooxygenase (Rahn and Hohmann 2009).

11.7 Antipruritic Effects

Pruritus, an unpleasant cutaneous sensation associated with an immediate desire to scratch, may be interpreted as one of the body's defense mechanisms (reviewed by Steinhoff et al. 2006). Pruriceptor stimulation conveys the transduced signal via histamine-positive mechano-insensitive C-fibers originating in DRG up to itch-selective units in lamina I of the spinal cord. Ascending signals project via posterior part of the ventromedial thalamic nucleus to finally reach the dorsal insular cortex (reviewed by Steinhoff et al. 2006). Inflammatory mediators released after disruption of cutaneous barrier function or UV radiation can stimulate sensory nerve endings (reviewed by Roosterman et al. 2006; Slominski and Wortsman 2000), and thus, induce pruritus (reviewed by Yosipovitch 2010). Peripheral administration of cannabinoid receptor agonists attenuated histamine-induced itch in humans (Dvorak et al. 2003). Recent study with the use of *N*-palmitoylethanolamine (cannabinoid-like amide acting through PPAR-α), which was added as a component of itch-relieving creams, alleviated pruritus in patients with atopic dermatitis, lichen simplex, and prurigo nodularis (Kircik 2010). These promising preliminary results suggest that new therapies targeting cannabinoid receptors may result in providing effective antipruritic medication in the future.

11.8 Conclusions

The recently described endocannabinoid system contributes to the abundant neuroendocrine activities of the skin. ECS participate in a number of pathophysiological processes in the skin and present there profound anti-inflammatory, antitumorigenic, antinociceptive, and antipruritic effects. ECS interact with two specific receptors which lead to synaptic inhibition of many neurotransmitter systems. Furthermore, ECS can interact with other receptors (opioid, serotonin, and TRPV) by nonspecific binding, modulating the release of other neurotransmitters and hormones. Thus, cutaneous ECS can participate in the regulation of local and systemic homeostasis (Figs. 1.1 and 1.2).

Chapter 12
Perspectives

Described as the body's largest organ, the skin is strategically located at the interface with the external environment where it has evolved to detect, integrate, and respond to a diverse range of stressors including UV radiation. Recent findings have established the skin as a peripheral neuroendocrine organ that is tightly networked to central stress axes (Fig. 1.2). This capability contributes to the maintenance of skin's and body's homeostasis. Specifically, epidermal and dermal cells produce and respond to classical stress neurotransmitters, neuropeptides, and hormones, and this production is modified by ultraviolet radiation and biological, chemical, and physical factors. Examples of potent epidermal products include biogenic amines (catecholamines, serotonin, and N-acetyl-serotonin) (Figs. 2.1–2.3), acetylcholine, melatonin and its metabolites (Figs. 2.5 and 3.1), proopiomelanocortin-derived ACTH, β-endorphin and MSH peptides, corticotropin-releasing factor and related urocortins (Figs. 5.1, 5.2, 7.3 and 7.4), corticosteroids and their precursor molecules, thyroid-related hormones (Fig. 9.1), opioids, and cannabinoids. The production of these molecules in the skin is hierarchical, following the algorithms of classical neuroendocrine axes (e.g., hypothalamic pituitary adrenal axis (HPA), hypothalamic–thyroid axis, serotoninergic/melatoninergic, catecholaminergic and cholinergic systems). The deregulation of these systems may be involved in the etiology of some skin diseases. These local neuroendocrine systems represent exquisite regulatory levels addressed at restricting the effect of noxious agents to preserve local and, consequently, global body's homeostasis and adapt to changing external environment. Furthermore, the skin-derived signals may also activate cutaneous sensory nerve endings to alert the brain on environment- or pathology-induced changes in the epidermal and dermal milieau, or alternatively, to activate other coordinating centers by spinal cord neurotransmission with or without brain's involvement (Fig. 1.1). Finally, the local neuroendocrine system will imprint resident and circulating immune cells to act as cellular messengers sent to other organs to coordinate responses to the changing environment (1.1).

A.T. Slominski et al., *Sensing the Environment: Regulation of Local and Global Homeostasis by the Skin's Neuroendocrine System*, Advances in Anatomy, Embryology and Cell Biology 212, DOI 10.1007/978-3-642-19683-6_12, © Springer-Verlag Berlin Heidelberg 2012

References

Abe M, Itoh MT, Miyata M, Ishikawa S, Sumi Y (1999) Detection of melatonin, its precursors and related enzyme activities in rabbit lens. Exp Eye Res 68:255–262

Aberg KM, Radek KA, Choi EH, Kim DK, Demerjian M, Hupe M, Kerbleski J, Gallo RL, Ganz T, Mauro T, Feingold KR, Elias PM (2007) Psychological stress downregulates epidermal antimicrobial peptide expression and increases severity of cutaneous infections in mice. J Clin Invest 117:3339–3349

Abrams DI, Couey P, Shade SB, Kelly ME, Benowitz NL (2011) Cannabinoid-opioid interaction in chronic pain. Clin Pharmacol Ther 90:844–851

Aguilera G, Rabadan-Diehl C, Nikodemova M (2001) Regulation of pituitary corticotropin releasing hormone receptors. Peptides 22:769–774

Akerman S, Kaube H, Goadsby PJ (2004) Anandamide acts as a vasodilator of dural blood vessels in vivo by activating TRPV1 receptors. Br J Pharmacol 142:1354–1360

Akhmetshina A, Dees C, Busch N, Beer J, Sarter K, Zwerina J, Zimmer A, Distler O, Schett G, Distler JH (2009) The cannabinoid receptor CB2 exerts antifibrotic effects in experimental dermal fibrosis. Arthritis Rheum 60:1129–1136

Albanesi C, Pastore S, Fanales-Belasio E, Girolomoni G (1998) Cetirizine and hydrocortisone differentially regulate ICAM-1 expression and chemokine release in cultured human keratinocytes. Clin Exp Allergy 28:101–109

Albro PW, Corbett JT, Schroeder JL (1994) Doubly allylic hydroperoxide formed in the reaction between sterol 5,7-dienes and singlet oxygen. Photochem Photobiol 60:310–315

Amaya F, Shimosato G, Kawasaki Y, Hashimoto S, Tanaka Y, Ji RR, Tanaka M (2006) Induction of CB1 cannabinoid receptor by inflammation in primary afferent neurons facilitates antihyperalgesic effect of peripheral CB1 agonist. Pain 124(1–2):175–183

Arck PC, Slominski A, Theoharides TC, Peters EM, Paus R (2006) Neuroimmunology of stress: skin takes center stage. J Invest Dermatol 126:1697–1704

Armario A (2010) Activation of the hypothalamic-pituitary-adrenal axis by addictive drugs: different pathways, common outcome. Trends Pharmacol Sci 31:318–325

Arredondo J, Nguyen VT, Chernyavsky AI, Bercovich D, Orr-Urtreger A, Kummer W, Lips K, Vetter DE, Grando SA (2002) Central role of alpha7 nicotinic receptor in differentiation of the stratified squamous epithelium. J Cell Biol 159:325–336

Arredondo J, Hall LL, Ndoye A, Chernyavsky AI, Jolkovsky DL, Grando SA (2003) Muscarinic acetylcholine receptors regulating cell cycle progression are expressed in human gingival keratinocytes. J Periodontal Res 38:79–89

A.T. Slominski et al., *Sensing the Environment: Regulation of Local and Global Homeostasis by the Skin's Neuroendocrine System*, Advances in Anatomy, Embryology and Cell Biology 212, DOI 10.1007/978-3-642-19683-6, © Springer-Verlag Berlin Heidelberg 2012

Ayoub MA, Couturier C, Lucas-Meunier E, Angers S, Fossier P, Bouvier M, Jockers R (2002) Monitoring of ligand-independent dimerization and ligand-induced conformational changes of melatonin receptors in living cells by bioluminescence resonance energy transfer. J Biol Chem 277:21522–21528

Azmitia EC (2001) Modern views on an ancient chemical: serotonin effects on cell proliferation, maturation, and apoptosis. Brain Res Bull 56:413–424

Azmitia EC (2007) Serotonin and brain: evolution, neuroplasticity, and homeostasis. Int Rev Neurobiol 77:31–56

Azmitia EC (2010) Evolution of serotonin: sunlight to suicide. In: Muller CP, Jacobs BL (eds) Handbook of the behavioral neurobiology of serotonin. Academic Press, Burlington, MA

Bangha E, Elsner P, Kistler GS (1996) Suppression of UV-induced erythema by topical treatment with melatonin (N-acetyl-5-methoxytryptamine). A dose response study. Arch Dermatol Res 288:522–526

Bangha E, Elsner P, Kistler GS (1997) Suppression of UV-induced erythema by topical treatment with melatonin (N-acetyl-5-methoxytryptamine). Influence of the application time point. Dermatology 195:248–252

Barr GA, Zadina JE (1999) Maturation of endomorphin-2 in the dorsal horn of the medulla and spinal cord of the rat. Neuroreport 10:3857–3860

Bartsch C, Bartsch H, Karasek M (2002) Melatonin in clinical oncology. Neuroendocrinol Lett 23 (Suppl 1):30–38

Beaulieu JM, Gainetdinov RR (2011) The physiology, signaling, and pharmacology of dopamine receptors. Pharmacol Rev 63:182–217

Becker-Andre M, Wiesenberg I, Schaeren-Wiemers N, Andre E, Missbach M, Saurat JH, Carlberg C (1994) Pineal gland hormone melatonin binds and activates an orphan of the nuclear receptor superfamily. J Biol Chem 269:28531–28534

Becklund BR, Severson KS, Vang SV, Deluca HF (2010) UV radiation suppresses experimental autoimmune encephalomyelitis independent of vitamin D production. Proc Natl Acad Sci USA 107:6418–6423

Belon PE (1985) UVA exposure and pituitary secretion. Variations of human lipotropin concentrations (beta LPH) after UVA exposure. Photochem Photobiol 42:327–329

Benton T, Lynch K, Dube B, Gettes DR, Tustin NB, Ping Lai J, Metzger DS, Blume J, Douglas SD, Evans DL (2010) Selective serotonin reuptake inhibitor suppression of HIV infectivity and replication. Psychosom Med 72:925–932

Bergmann M, Sautner T (2002) Immunomodulatory effects of vasoactive catecholamines. Wien Klin Wochenschr 114:752–761

Bergquist J, Josefsson E, Tarkowski A, Ekman R, Ewing A (1997) Measurements of catecholamine-mediated apoptosis of immunocompetent cells by capillary electrophoresis. Electrophoresis 18:1760–1766

Berwick M, Armstrong BK, Ben-Porat L, Fine J, Kricker A, Eberle C, Barnhill R (2005) Sun exposure and mortality from melanoma. J Natl Cancer Inst 97:195–199

Besedovsky HO, Rey AD (2007) Physiology of psychoneuroimmunology: a personal view. Brain Behav Immun 21:34–44

Besser MJ, Ganor Y, Levite M (2005) Dopamine by itself activates either D2, D3 or D1/D5 dopaminergic receptors in normal human T-cells and triggers the selective secretion of either IL-10, TNFalpha or both. J Neuroimmunol 169:161–171

Betten A, Dahlgren C, Hermodsson S, Hellstrand K (2001) Serotonin protects NK cells against oxidatively induced functional inhibition and apoptosis. J Leukoc Biol 70:65–72

Bigliardi PL, Bigliardi-Qi M, Buechner S, Rufli T (1998) Expression of mu-opiate receptor in human epidermis and keratinocytes. J Invest Dermatol 111:297–301

Bigliardi PL, Tobin DJ, Gaveriaux-Ruff C, Bigliardi-Qi M (2009) Opioids and the skin—where do we stand? Exp Dermatol 18:424–430

Bikle DD (2010) Vitamin D and the skin. J Bone Miner Metab 28:117–130

Bikle DD (2011a) Vitamin D metabolism and function in the skin. Mol Cell Endocrinol 347:80–89

Bikle DD (2011b) The vitamin D receptor: a tumor suppressor in skin. Discov Med 11:7–17

Bikle DD (2011c) Vitamin D: an ancient hormone. Exp Dermatol 20:7–13

Billoni N, Buan B, Gautier B, Gaillard O, Mahe YF, Bernard BA (2000) Thyroid hormone receptor beta1 is expressed in the human hair follicle. Br J Dermatol 142:645–652

Biro T, Toth BI, Hasko G, Paus R, Pacher P (2009) The endocannabinoid system of the skin in health and disease: novel perspectives and therapeutic opportunities. Trends Pharmacol Sci 30:411–420

Bissonnette EY, Befus AD (1997) Anti-inflammatory effect of beta 2-agonists: inhibition of TNF-alpha release from human mast cells. J Allergy Clin Immunol 100:825–831

Blalock JE, Smith EM (2007) Conceptual development of the immune system as a sixth sense. Brain Behav Immun 21:23–33

Blau N, Van Spronsen FJ, Levy HL (2010) Phenylketonuria. Lancet 376:1417–1427

Blazquez C, Carracedo A, Barrado L, Real PJ, Fernandez-Luna JL, Velasco G, Malumbres M, Guzman M (2006) Cannabinoid receptors as novel targets for the treatment of melanoma. FASEB J 20:2633–2635

Bodo E, Kromminga A, Biro T, Borbiro I, Gaspar E, Zmijewski MA, Van Beek N, Langbein L, Slominski AT, Paus R (2009) Human female hair follicles are a direct, nonclassical target for thyroid-stimulating hormone. J Invest Dermatol 129:1126–1139

Bodo E, Kany B, Gaspar E, Knuver J, Kromminga A, Ramot Y, Biro T, Tiede S, Van Beek N, Poeggeler B, Meyer KC, Wenzel BE, Paus R (2010) Thyroid-stimulating hormone, a novel, locally produced modulator of human epidermal functions, is regulated by thyrotropin-releasing hormone and thyroid hormones. Endocrinology 151:1633–1642

Bolognia J, Jorizzo Jl, Rapini RP (2008) Dermatology. Mosby—Elsevier, Philadelphia, PA

Borzsei R, Pozsgai G, Bagoly T, Elekes K, Pinter E, Szolcsanyi J, Helyes Z (2008) Inhibitory action of endomorphin-1 on sensory neuropeptide release and neurogenic inflammation in rats and mice. Neuroscience 152:82–88

Bouatia-Naji N, Bonnefond A, Cavalcanti-Proenca C, Sparso T, Holmkvist J, Marchand M et al (2009) A variant near MTNR1B is associated with increased fasting plasma glucose levels and type 2 diabetes risk. Nat Genet 41:89–94

Branchek TA, Mawe GM, Gershon MD (1988) Characterization and localization of a peripheral neural 5-hydroxytryptamine receptor subtype (5-HT1P) with a selective agonist, 3H-5-hydroxyindalpine. J Neurosci 8:2582–2595

Brozyna AA, Jozwicki W, Janjetovic Z, Slominski AT (2011) Expression of vitamin D receptor decreases during progression of pigmented skin lesions. Hum Pathol 42:618–631

Bubenik GA (2002) Gastrointestinal melatonin: localization, function, and clinical relevance. Dig Dis Sci 47:2336–2348

Buchli R, Ndoye A, Arredondo J, Webber RJ, Grando SA (2001) Identification and characterization of muscarinic acetylcholine receptor subtypes expressed in human skin melanocytes. Mol Cell Biochem 228:57–72

Bujalska I, Shimojo M, Howie A, Stewart PM (1997) Human 11 beta-hydroxysteroid dehydrogenase: studies on the stably transfected isoforms and localization of the type 2 isozyme within renal tissue. Steroids 62:77–82

Bujalska IJ, Walker EA, Hewison M, Stewart PM (2002) A switch in dehydrogenase to reductase activity of 11 beta-hydroxysteroid dehydrogenase type 1 upon differentiation of human omental adipose stromal cells. J Clin Endocrinol Metab 87:1205–1210

Cahill GM, Besharse JC (1989) Retinal melatonin is metabolized within the eye of Xenopus laevis. Proc Natl Acad Sci USA 86:1098–1102

Capsoni S, Viswanathan M, De Oliveira AM, Saavedra JM (1994) Characterization of melatonin receptors and signal transduction system in rat arteries forming the circle of Willis. Endocrinology 135:373–378

Carlberg C, Hooft Van Huijsduijnen R, Staple JK, Delamarter JF, Becker-Andre M (1994) RZRs, a new family of retinoid-related orphan receptors that function as both monomers and homodimers. Mol Endocrinol 8:757–770

Carlton SM, Coggeshall RE (1997) Immunohistochemical localization of 5-HT2A receptors in peripheral sensory axons in rat glabrous skin. Brain Res 763:271–275

Carrier Y, Ma HL, Ramon HE, Napierata L, Small C, O'toole M, Young DA, Fouser LA, Nickerson-Nutter C, Collins M, Dunussi-Joannopoulos K, Medley QG (2011) Inter-regulation of Th17 cytokines and the il-36 cytokines in vitro and in vivo: implications in psoriasis pathogenesis. J Invest Dermatol 131:2428–2437

Carrillo-Vico A, Calvo JR, Abreu P, Lardone PJ, Garcia-Maurino S, Reiter RJ, Guerrero JM (2004) Evidence of melatonin synthesis by human lymphocytes and its physiological significance: possible role as intracrine, autocrine, and/or paracrine substance. FASEB J 18:537–539

Casanova ML, Blazquez C, Martinez-Palacio J, Villanueva C, Fernandez-Acenero MJ, Huffman JW, Jorcano JL, Guzman M (2003) Inhibition of skin tumor growth and angiogenesis in vivo by activation of cannabinoid receptors. J Clin Invest 111:43–50

Chakraborty AK, Funasaka Y, Slominski A, Bolognia J, Sodi S, Ichihashi M, Pawelek JM (1999) UV light and MSH receptors. Ann N Y Acad Sci 885:100–116

Chen J, Hoffman BB, Isseroff RR (2002) Beta-adrenergic receptor activation inhibits keratinocyte migration via a cyclic adenosine monophosphate-independent mechanism. J Invest Dermatol 119:1261–1268

Chernyavsky AI, Arredondo J, Marubio LM, Grando SA (2004a) Differential regulation of keratinocyte chemokinesis and chemotaxis through distinct nicotinic receptor subtypes. J Cell Sci 117:5665–5679

Chernyavsky AI, Arredondo J, Wess J, Karlsson E, Grando SA (2004b) Novel signaling pathways mediating reciprocal control of keratinocyte migration and wound epithelialization through M3 and M4 muscarinic receptors. J Cell Biol 166:261–272

Chernyavsky AI, Arredondo J, Piser T, Karlsson E, Grando SA (2008) Differential coupling of M1 muscarinic and alpha7 nicotinic receptors to inhibition of pemphigus acantholysis. J Biol Chem 283:3401–3408

Chesnokova V, Melmed S (2002) Minireview: neuro-immuno-endocrine modulation of the hypothalamic-pituitary-adrenal (HPA) axis by gp130 signaling molecules. Endocrinology 143:1571–1574

Chignell CF, Kukielczak BM, Sik RH, Bilski PJ, He YY (2006) Ultraviolet al. sensitivity in Smith-Lemli-Opitz syndrome: possible involvement of cholesta-5,7,9(11)-trien-3 beta-ol. Free Radic Biol Med 41:339–346

Chizhmakov I, Yudin Y, Mamenko N, Prudnikov I, Tamarova Z, Krishtal O (2005) Opioids inhibit purinergic nociceptors in the sensory neurons and fibres of rat via a G protein-dependent mechanism. Neuropharmacology 48:639–647

Chrousos GP (1995) The hypothalamic-pituitary-adrenal axis and immune-mediated inflammation. N Engl J Med 332:1351–1362

Chrousos GP, Gold PW (1992) The concepts of stress and stress system disorders. Overview of physical and behavioral homeostasis. JAMA 267:1244–1252

Cianfarani F, Baldini E, Cavalli A, Marchioni E, Lembo L, Teson M, Persechino S, Zambruno G, Ulisse S, Odorisio T, D'armiento M (2010) TSH receptor and thyroid-specific gene expression in human skin. J Invest Dermatol 130:93–101

Cirillo N, Prime SS (2011) Keratinocytes synthesize and activate cortisol. J Cell Biochem 112:1499–1505

Cloez-Tayarani I, Changeux JP (2007) Nicotine and serotonin in immune regulation and inflammatory processes: a perspective. J Leukoc Biol 81:599–606

Coburn SP, Slominski A, Mahuren JD, Wortsman J, Hessle L, Millan JL (2003) Cutaneous metabolism of vitamin B-6. J Invest Dermatol 120:292–300

Connell K, Bolton N, Olsen D, Piomelli D, Hohmann AG (2006) Role of the basolateral nucleus of the amygdala in endocannabinoid-mediated stress-induced analgesia. Neurosci Lett 397:180–184

Contreras-Jurado C, Garcia-Serrano L, Gomez-Ferreria M, Costa C, Paramio JM, Aranda A (2011) The thyroid hormone receptors as modulators of skin proliferation and inflammation. J Biol Chem 286:24079–24088

Costa B, Pini S, Gabelloni P, Da Pozzo E, Abelli M, Lari L, Preve M, Lucacchini A, Cassano GB, Martini C (2009) The spontaneous Ala147Thr amino acid substitution within the translocator protein influences pregnenolone production in lymphomonocytes of healthy individuals. Endocrinology 150:5438–5445

Cotecchia S (2010) The alpha1-adrenergic receptors: diversity of signaling networks and regulation. J Recept Signal Transduct Res 30:410–419

Dai J, Ram PT, Yuan L, Spriggs LL, Hill SM (2001) Transcriptional repression of RORalpha activity in human breast cancer cells by melatonin. Mol Cell Endocrinol 176:111–120

Davies E, MacKenzie SM (2003) Extra-adrenal production of corticosteroids. Clin Exp Pharmacol Physiol 30:437–445

Davis SC, Clark S, Hayes JR, Green TL, Gruetter CA (2011) Up-regulation of histidine decarboxylase expression and histamine content in B16F10 murine melanoma cells. Inflamm Res 60:55–61

De Fabiani E, Caruso D, Cavaleri M, Galli Kienle M, Galli G (1996) Cholesta-5,7,9(11)-trien-3 beta-ol found in plasma of patients with Smith-Lemli-Opitz syndrome indicates formation of sterol hydroperoxide. J Lipid Res 37:2280–2287

Denzer N, Vogt T, Reichrath J (2011) Vitamin D receptor (VDR) polymorphisms and skin cancer: a systematic review. Dermatoendocrinology 3:205–210

Dijkstra D, Leurs R, Chazot P, Shenton FC, Stark H, Werfel T, Gutzmer R (2007) Histamine downregulates monocyte CCL2 production through the histamine H4 receptor. J Allergy Clin Immunol 120:300–307

Ding Z, Jiang M, Li S, Zhang Y (1995) Vascular barrier-enhancing effect of an endogenous beta-adrenergic agonist. Inflammation 19:1–8

Do Rego JL, Seong JY, Burel D, Leprince J, Luu-The V, Tsutsui K, Tonon MC, Pelletier G, Vaudry H (2009) Front Neuroendocrinol. 30:259–301

Doshi DN, Blyumin ML, Kimball AB (2008) Cutaneous manifestations of thyroid disease. Clin Dermatol 26:283–287

Draper N, Stewart PM (2005) 11beta-hydroxysteroid dehydrogenase and the pre-receptor regulation of corticosteroid hormone action. J Endocrinol 186:251–271

Dreher F, Gabard B, Schwindt DA, Maibach HI (1998) Topical melatonin in combination with vitamins E and C protects skin from ultraviolet-induced erythema: a human study in vivo. Br J Dermatol 139:332–339

Dreher F, Denig N, Gabard B, Schwindt DA, Maibach HI (1999) Effect of topical antioxidants on UV-induced erythema formation when administered after exposure. Dermatol 198:52–55

Drummond PD, Skipworth S, Finch PM (1996) alpha 1-adrenoceptors in normal and hyperalgesic human skin. Clin Sci (Lond) 91:73–77

Dubocovich ML, Markowska M (2005) Functional MT1 and MT2 melatonin receptors in mammals. Endocrine 27:101–110

Dubocovich ML, Rivera-Bermudez MA, Gerdin MJ, Masana MI (2003) Molecular pharmacology, regulation and function of mammalian melatonin receptors. Front Biosci 8:d1093–d1108

Dubocovich ML, Delagrange P, Krause DN, Sugden D, Cardinali DP, Olcese J (2010) International Union of Basic and Clinical Pharmacology. LXXV. Nomenclature, classification, and pharmacology of G protein-coupled melatonin receptors. Pharmacol Rev 62:343–380

Dumont M, Luu-The V, Dupont E, Pelletier G, Labrie F (1992) Characterization, expression, and immunohistochemical localization of 3 beta-hydroxysteroid dehydrogenase/delta 5-delta 4 isomerase in human skin. J Invest Dermatol 99:415–421

Dvorak M, Watkinson A, Mcglone F, Rukwied R (2003) Histamine induced responses are attenuated by a cannabinoid receptor agonist in human skin. Inflamm Res 52:238–245

Eisenhofer G, Tian H, Holmes C, Matsunaga J, Roffler-Tarlov S, Hearing VJ (2003) Tyrosinase: a developmentally specific major determinant of peripheral dopamine. FASEB J 17:1248–1255

Ellerhorst JA, Sendi-Naderi A, Johnson MK, Cooke CP, Dang SM, Diwan AH (2006) Human melanoma cells express functional receptors for thyroid-stimulating hormone. Endocr Relat Cancer 13:1269–1277

Ellerhorst JA, Diwan AH, Dang SM, Uffort DG, Johnson MK, Cooke CP, Grimm EA (2010) Promotion of melanoma growth by the metabolic hormone leptin. Oncol Rep 23:901–907

Ermak G, Slominski A (1997) Production of POMC, CRH-, ACTH- and a-MSH- receptor mRNA and expression of tyrosinase gene in relation to hair cycle and dexamethasone treatment in the C57BL/6 mouse skin. J Invest Dermatol 108:160–167

Falck B, Bendsoe N, Ronquist G (2004) Mediated exodus of L-dopa from human epidermal Langerhans cells. Amino Acids 26:133–138

Fallahzadeh MK, Namazi MR (2009) Opioid-mediated immunosuppression as a novel mechanism for the immunomodulatory effect of ultraviolet radiation. Indian J Dermatol Venereol Leprol 75:622–623

Fazal N, Slominski A, Choudhry MA, Wei ET, Sayeed MM (1998) Effect of CRF and related peptides on calcium signaling in human and rodent melanoma cells. FEBS Lett 435:187–190

Felig P, Frohman LA (2001) Endocrinology and metabolism. McGraw-Hill, New York, NY

Feng K, Zhang RY, Wu LZ, Tu B, Peng ML, Zhang LP, Zhao D, Tung CH (2006) Photooxidation of olefins under oxygen in platinum(II) complex-loaded mesoporous molecular sieves. J Am Chem Soc 128:14685–14690

Fichna J, Janecka A, Costentin J, Do Rego JC (2007) The endomorphin system and its evolving neurophysiological role. Pharmacol Rev 59:88–123

Field S, Newton-Bishop JA (2011) Melanoma and vitamin D. Mol Oncol 5:197–214

Finn DP (2010) Endocannabinoid-mediated modulation of stress responses: physiological and pathophysiological significance. Immunobiol 215:629–646

Finocchiaro LM, Nahmod VE, Launay JM (1991) Melatonin biosynthesis and metabolism in peripheral blood mononuclear leucocytes. Biochem J 280(Pt 3):727–731

Fischer J, Bouadjar B, Heilig R, Huber M, Lefevre C, Jobard F, Macari F, Bakija-Konsuo A, Ait-Belkacem F, Weissenbach J, Lathrop M, Hohl D, Prud'homme JF (2001) Mutations in the gene encoding SLURP-1 in Mal de Meleda. Hum Mol Genet 10:875–880

Fischer TW, Scholz G, Knoll B, Hipler UC, Elsner P (2002) Melatonin suppresses reactive oxygen species in UV-irradiated leukocytes more than vitamin C and trolox. Skin Pharmacol Appl Skin Physiol 15:367–373

Fischer TW, Burmeister G, Schmidt HW, Elsner P (2004) Melatonin increases anagen hair rate in women with androgenetic alopecia or diffuse alopecia: results of a pilot randomized controlled trial. Br J Dermatol 150:341–345

Fischer TW, Sweatman TW, Semak I, Sayre RM, Wortsman J, Slominski A (2006a) Constitutive and UV-induced metabolism of melatonin in keratinocytes and cell-free systems. FASEB J 20:1564–1566

Fischer TW, Zbytek B, Sayre RM, Apostolov EO, Basnakian AG, Sweatman TW, Wortsman J, Elsner P, Slominski A (2006b) Melatonin increases survival of HaCaT keratinocytes by suppressing UV-induced apoptosis. J Pineal Res 40:18–26

Fischer TW, Zmijewski MA, Zbytek B, Sweatman TW, Slominski RM, Wortsman J, Slominski A (2006c) Oncostatic effects of the indole melatonin and expression of its cytosolic and nuclear receptors in cultured human melanoma cell lines. Int J Oncol 29:665–672

Fischer TW, Slominski A, Tobin DJ, Paus R (2008a) Melatonin and the hair follicle. J Pineal Res 44:1–15

Fischer TW, Slominski A, Zmijewski MA, Reiter RJ, Paus R (2008b) Melatonin as a major skin protectant: from free radical scavenging to DNA damage repair. Exp Dermatol 17:713–730

Fischer TW, Zmijewski MA, Wortsman J, Slominski A (2008c) Melatonin maintains mitochondrial membrane potential and attenuates activation of initiator (casp-9) and effector caspases (casp-3/casp-7) and PARP in UVR-exposed HaCaT keratinocytes. J Pineal Res 44:397–407

Fitzpatrick TB, Wolff K, Freedberg IM, Austen KF (1993) Dermatology in general medicine. McGraw Hill, New York

Fitzsimons C, Engel N, Policastro L, Duran H, Molinari B, Rivera E (2002) Regulation of phospholipase C activation by the number of H(2) receptors during Ca(2+)-induced differentiation of mouse keratinocytes. Biochem Pharmacol 63:1785–1796

Fromy B, Sigaudo-Roussel D, Gaubert-Dahan ML, Rousseau P, Abraham P, Benzoni D, Berrut G, Saumet JL (2010) Aging-associated sensory neuropathy alters pressure-induced vasodilation in humans. J Invest Dermatol 130:849–855

Fujii E, Irie K, Uchida Y, Tsukahara F, Muraki T (1994) Possible role of nitric oxide in 5-hydroxytryptamine-induced increase in vascular permeability in mouse skin. Naunyn Schmiedebergs Arch Pharmacol 350:361–364

Fujimoto S, Komine M, Karakawa M, Uratsuji H, Kagami S, Tada Y, Saeki H, Ohtsuki M, Tamaki K (2011) Histamine differentially regulates the production of Th1 and Th2 chemokines by keratinocytes through histamine H1 receptor. Cytokine 54:191–199

Funasaka Y, Sato H, Chakraborty AK, Ohashi A, Chrousos GP, Ichihashi M (1999) Expression of proopiomelanocortin, corticotropin-releasing hormone (CRH), and CRH receptor in melanoma cells, nevus cells, and normal human melanocytes. J Investig Dermatol Symp Proc 4:105–109

Fuziwara S, Suzuki A, Inoue K, Denda M (2005) Dopamine D2-like receptor agonists accelerate barrier repair and inhibit the epidermal hyperplasia induced by barrier disruption. J Invest Dermatol 125:783–789

Gambichler T, Bader A, Vojvodic M, Bechara FG, Sauermann K, Altmeyer P, Hoffmann K (2002) Impact of UVA exposure on psychological parameters and circulating serotonin and melatonin. BMC Dermatol 2:6

Garssen J, De Gruijl F, Mol D, De Klerk A, Roholl P, Van Loveren H (2001) UVA exposure affects UVB and cis-urocanic acid-induced systemic suppression of immune responses in Listeria monocytogenes-infected Balb/c mice. Photochem Photobiol 73:432–438

Gaspar E, Hardenbicker C, Bodo E, Wenzel B, Ramot Y, Funk W, Kromminga A, Paus R (2010) Thyrotropin releasing hormone (TRH): a new player in human hair-growth control. FASEB J 24:393–403

Gaspar E, Nguyen-Thi KT, Hardenbicker C, Tiede S, Plate C, Bodo E, Knuever J, Funk W, Biro T, Paus R (2011) Thyrotropin-releasing hormone selectively stimulates human hair follicle pigmentation. J Invest Dermatol 131:2368–2377

Gaudet SJ, Slominski A, Etminan M, Pruski D, Paus R, MaA N (1993) Identification and characterization of two isozymic forms of arylamine N-acetyltransferase in Syrian hamster skin. J Invest Dermatol 101:660–665

Ghoghawala SY, Mannis MJ, Pullar CE, Rosenblatt MI, Isseroff RR (2008) Beta2-adrenergic receptor signaling mediates corneal epithelial wound repair. Invest Ophthalmol Vis Sci 49:1857–1863

Gibbins IL (1992) Vasoconstrictor, vasodilator and pilomotor pathways in sympathetic ganglia of guinea-pigs. Neurosci 47:657–672

Gillbro JM, Marles LK, Hibberts NA, Schallreuter KU (2004) Autocrine catecholamine biosynthesis and the beta-adrenoceptor signal promote pigmentation in human epidermal melanocytes. J Invest Dermatol 123:346–353

Giustizieri ML, Albanesi C, Fluhr J, Gisondi P, Norgauer J, Girolomoni G (2004) H1 histamine receptor mediates inflammatory responses in human keratinocytes. J Allergy Clin Immunol 114:1176–1182

Gombart AF, Borregaard N, Koeffler HP (2005) Human cathelicidin antimicrobial peptide (CAMP) gene is a direct target of the vitamin D receptor and is strongly up-regulated in myeloid cells by 1,25-dihydroxyvitamin D3. FASEB J 19:1067–1077

Grace MS, Cahill GM, Besharse JC (1991) Melatonin deacetylation: retinal vertebrate class distribution and *Xenopus laevis* tissue distribution. Brain Res 559:56–63

Graef S, Schonknecht P, Sabri O, Hegerl U (2011) Cholinergic receptor subtypes and their role in cognition, emotion, and vigilance control: an overview of preclinical and clinical findings. Psychopharmacol (Berl) 215:205–229

Grammatopoulos DK, Chrousos GP (2002) Functional characteristics of CRH receptors and potential clinical applications of CRH-receptor antagonists. Trends Endocrinol Metab 13:436–444

Grammatopoulos DK, Dai Y, Randeva HS, Levine MA, Karteris E, Easton AJ, Hillhouse EW (1999) A novel spliced variant of the type 1 corticotropin-releasing hormone receptor with a deletion in the seventh transmembrane domain present in the human pregnant term myometrium and fetal membranes. Mol Endocrinol 13:2189–2202

Grando SA (1997) Biological functions of keratinocyte cholinergic receptors. J Investig Dermatol Symp Proc 2:41–48

Grando SA (2006) Cholinergic control of epidermal cohesion. Exp Dermatol 15:265–282

Grando SA, Kist DA, Qi M, Dahl MV (1993) Human keratinocytes synthesize, secrete, and degrade acetylcholine. J Invest Dermatol 101:32–36

Grando SA, Zelickson BD, Kist DA, Weinshenker D, Bigliardi PL, Wendelschafer-Crabb G, Kennedy WR, Dahl MV (1995) Keratinocyte muscarinic acetylcholine receptors: immunolocalization and partial characterization. J Invest Dermatol 104:95–100

Grando SA, Horton RM, Mauro TM, Kist DA, Lee TX, Dahl MV (1996) Activation of keratinocyte nicotinic cholinergic receptors stimulates calcium influx and enhances cell differentiation. J Invest Dermatol 107:412–418

Grando SA, Pittelkow MR, Schallreuter KU (2006) Adrenergic and cholinergic control in the biology of epidermis: physiological and clinical significance. J Invest Dermatol 126:1948–1965

Gschwandtner M, Purwar R, Wittmann M, Baumer W, Kietzmann M, Werfel T, Gutzmer R (2008) Histamine upregulates keratinocyte MMP-9 production via the histamine H1 receptor. J Invest Dermatol 128:2783–2791

Gschwandtner M, Mommert S, Kother B, Werfel T, Gutzmer R (2011) The histamine H4 receptor is highly expressed on plasmacytoid dendritic cells in psoriasis and histamine regulates their cytokine production and migration. J Invest Dermatol 131:1668–1676

Guindon J, Hohmann AG (2009) The endocannabinoid system and pain. CNS Neurol Disord Drug Targets 8:403–421

Guryev O, Carvalho RA, Usanov S, Gilep A, Estabrook RW (2003) A pathway for the metabolism of vitamin D3: unique hydroxylated metabolites formed during catalysis with cytochrome P450scc (CYP11A1). Proc Natl Acad Sci USA 100:14754–14759

Gutzmer R, Mommert S, Gschwandtner M, Zwingmann K, Stark H, Werfel T (2009) The histamine H4 receptor is functionally expressed on T(H)2 cells. J Allergy Clin Immunol 123:619–625

Haak-Frendscho M, Darvas Z, Hegyesi H, Karpati S, Hoffman RL, Laszlo V, Bencsath M, Szalai C, Furesz J, Timar J, Bata-Csorgo Z, Szabad G, Pivarcsi A, Pallinger E, Kemeny L, Horvath A, Dobozy A, Falus A (2000) Histidine decarboxylase expression in human melanoma. J Invest Dermatol 115:345–352

Hannen RF, Michael AE, Jaulim A, Bhogal R, Burrin JM, Philpott MP (2011) Steroid synthesis by primary human keratinocytes; implications for skin disease. Biochem Biophys Res Commun 404:62–67

Hara MR, Kovacs JJ, Whalen EJ, Rajagopal S, Strachan RT, Grant W, Towers AJ, Williams B, Lam CM, Xiao K, Shenoy SK, Gregory SG, Ahn S, Duckett DR, Lefkowitz RJ (2011) A stress response pathway regulates DNA damage through beta(2)-adrenoreceptors and beta-arrestin-1. Nature 477:349–353

Harada K, Ohashi K, Fujimura A, Kumagai Y, Ebihara A (1996) Effect of alpha 1-adrenoceptor antagonists, prazosin and urapidil, on a finger skin vasoconstrictor response to cold stimulation. Eur J Clin Pharmacol 49:371–375

Hardeland R, Backhaus C, Fadavi A (2007) Reactions of the NO redox forms NO+, *NO and HNO (protonated NO–) with the melatonin metabolite N1-acetyl-5-methoxykynuramine. J Pineal Res 43:382–388

Hardeland R, Tan DX, Reiter RJ (2009) Kynuramines, metabolites of melatonin and other indoles: the resurrection of an almost forgotten class of biogenic amines. J Pineal Res 47:109–126

Hardeland R, Cardinali DP, Srinivasan V, Spence DW, Brown GM, Pandi-Perumal SR (2011) Melatonin–a pleiotropic, orchestrating regulator molecule. Prog Neurobiol 93:350–384

Harrington LE, Mangan PR, Weaver CT (2006) Expanding the effector CD4 T-cell repertoire: the Th17 lineage. Curr Opin Immunol 18:349–356

Harrington CR, Beswick TC, Graves M, Jacobe HT, Harris TS, Kourosh S, Devous Sr MD, Adinoff B (2012) Activation of the mesostriatal reward pathway with exposure to ultraviolet radiation (UVR) vs. sham UVR in frequent tanners: a pilot study. Addict Biol. 17:680–6

Hassan AH, Pzewlocki R, Herz A, Stein C (1992) Dynorphin, a preferential ligand for kappa-opioid receptors, is present in nerve fibers and immune cells within inflamed tissue of the rat. Neurosci Lett 140:85–88

Hasse S, Chernyavsky AI, Grando SA, Paus R (2007) The M4 muscarinic acetylcholine receptor plays a key role in the control of murine hair follicle cycling and pigmentation. Life Sci 80:2248–2252

Hein L (2006) Adrenoceptors and signal transduction in neurons. Cell Tissue Res 326:541–551

Hemley CF, Mccluskey A, Keller PA (2007) Corticotropin releasing hormone–a GPCR drug target. Curr Drug Targets 8:105–115

Hickman AB, Namboodiri MA, Klein DC, Dyda F (1999) The structural basis of ordered substrate binding by serotonin N-acetyltransferase: enzyme complex at 1.8 A resolution with a bisubstrate analog. Cell 97:361–369

Hill SM, Frasch T, Xiang S, Yuan L, Duplessis T, Mao L (2009) Molecular mechanisms of melatonin anticancer effects. Integr Cancer Ther 8:337–346

Hillhouse EW, Grammatopoulos DK (2006) The molecular mechanisms underlying the regulation of the biological activity of corticotropin-releasing hormone receptors: implications for physiology and pathophysiology. Endocr Rev 27:260–286

Hillhouse EW, Randeva H, Ladds G, Grammatopoulos D (2002) Corticotropin-releasing hormone receptors. Biochem Soc Trans 30:428–432

Hirata F, Hayaishi O, Tokuyama T, Seno S (1974) In vitro and in vivo formation of two new metabolites of melatonin. J Biol Chem 249:1311–1313

Holick MF (2003) Vitamin D: a millenium perspective. J Cell Biochem 88:296–307

Holick MF (2008) Sunlight, UV-radiation, vitamin D and skin cancer: how much sunlight do we need? Adv Exp Med Biol 624:1–15

Holick MF, Clark MB (1978) The photobiogenesis and metabolism of vitamin D. Fed Proc 37:2567–2574

Holick MF, Tian XQ, Allen M (1995) Evolutionary importance for the membrane enhancement of the production of vitamin D3 in the skin of poikilothermic animals. Proc Natl Acad Sci USA 92:3124–3126

Holtzmann H, Altmeyer P, Schultz-Amling W (1982) Der Einfluss Ultravioletter Strahlen auf die Hypothalamus-Hypophysenachse des Menschen. Acta Dermatol 8:119–123

Holtzmann HP, Altmeyer L, Stohr CGN (1983) Die Beeinflussung des alpha-MSH durch UVA-Bestrahlunger der Haut—ein Funktionstest. Hautarzt 34:294–297

Hou P, Liu D, Ji M, Liu Z, Engles JM, Wahl RL, Xing M (2009) Induction of thyroid gene expression and radioiodine uptake in melanoma cells: novel therapeutic implications. PLoS One 4:e6200

Howe J, Costantino R, Slominski A (1991) On the putative mechanism of induction and regulation of melanogenesis by L-tyrosine. Acta Derm Venereol 71:150–152

Hoyer D, Hannon JP, Martin GR (2002) Molecular, pharmacological and functional diversity of 5-HT receptors. Pharmacol Biochem Behav 71:533–554

Hsueh CM, Chen SF, Lin RJ, Chao HJ (2002) Cholinergic and serotonergic activities are required in triggering conditioned NK cell response. J Neuroimmunol 123:102–111

Hu DN (2000) Regulation of growth and melanogenesis of uveal melanocytes. Pigment Cell Res 13(Suppl 8):81–86

Hu DN, Woodward DF, Mccormick SA (2000) Influence of autonomic neurotransmitters on human uveal melanocytes in vitro. Exp Eye Res 71:217–224

Huang MP, Rodgers KA, O'mara R, Mehta M, Abuzahra HS, Tannenbaum AD, Persons K, Holick MF, Safer JD (2011) The thyroid hormone degrading type 3 deiodinase is the primary deiodinase active in murine epidermis. Thyroid 21:1263–1268

Huttunen M, Hyttinen M, Nilsson G, Butterfield JH, Horsmanheimo M, Harvima IT (2001) Inhibition of keratinocyte growth in cell culture and whole skin culture by mast cell mediators. Exp Dermatol 10:184–192

Ibraheem M, Galbraith H, Scaife J, Ewen S (1994) Growth of secondary hair follicles of the Cashmere goat in vitro and their response to prolactin and melatonin. J Anat 185(Pt 1):135–142

Ibrahim MM, Porreca F, Lai J, Albrecht PJ, Rice FL, Khodorova A, Davar G, Makriyannis A, Vanderah TW, Mata HP, Malan TP Jr (2005) CB2 cannabinoid receptor activation produces antinociception by stimulating peripheral release of endogenous opioids. Proc Natl Acad Sci USA 102:3093–3098

Ingordo V, Gentile C, Iannazzone SS, Cusano F, Naldi L (2011) Vitiligo and autoimmunity: an epidemiological study in a representative sample of young Italian males. J Eur Acad Dermatol Venereol 25:105–109

Irving AJ, Rae MG, Coutts AA (2002) Cannabinoids in the brain. Scientific World J 2:632–648

Ishikawa T, Kanda N, Hau CS, Tada Y, Watanabe S (2009) Histamine induces human beta-defensin-3 production in human keratinocytes. J Dermatol Sci 56:121–127

Ito N, Ito T, Kromminga A, Bettermann A, Takigawa M, Kees F, Straub RH, Paus R (2005) Human hair follicles display a functional equivalent of the hypothalamic-pituitary-adrenal axis and synthesize cortisol. FASEB J 19:1332–1334

Itoh MT, Shinozawa T, Sumi Y (1999) Circadian rhythms of melatonin-synthesizing enzyme activities and melatonin levels in planarians. Brain Res 830:165–173

Itoi K, Jiang YQ, Iwasaki Y, Watson SJ (2004) Regulatory mechanisms of corticotropin-releasing hormone and vasopressin gene expression in the hypothalamus. J Neuroendocrinol 16:348–355

Iyengar B (1989) Modulation of melanocytic activity by acetylcholine. Acta Anat (Basel) 136:139–141

Jangi SM, Diaz-Perez JL, Ochoa-Lizarralde B, Martin-Ruiz I, Asumendi A, Perez-Yarza G, Gardeazabal J, Diaz-Ramon JL, Boyano MD (2006) H1 histamine receptor antagonists induce genotoxic and caspase-2-dependent apoptosis in human melanoma cells. Carcinogenesis 27:1787–1796

Janjetovic Z, Zmijewski MA, Tuckey RC, Deleon DA, Nguyen MN, Pfeffer LM, Slominski AT (2009) 20-Hydroxycholecalciferol, product of vitamin D3 hydroxylation by P450scc, decreases NF-kappaB activity by increasing IkappaB alpha levels in human keratinocytes. PLoS One 4:e5988

Janjetovic Z, Tuckey RC, Nguyen MN, Thorpe EM Jr, Slominski AT (2010) 20,23-dihydroxyvitamin D3, novel P450scc product, stimulates differentiation and inhibits proliferation and NF-kappaB activity in human keratinocytes. J Cell Physiol 223:36–48

Janjetovic Z, Brozyna AA, Tuckey RC, Kim TK, Nguyen MN, Jozwicki W, Pfeffer SR, Pfeffer LM, Slominski AT (2011) High basal NF-κB activity in nonpigmented melanoma cells is associated with an enhanced sensitivity to vitamin D3 derivatives. Br J Cancer 105:1874–1884

Jankowska E, Schomburg ED (1998) A leu-enkephalin depresses transmission from muscle and skin non-nociceptors to first-order feline spinal neurones. J Physiol 510(Pt 2):513–525

Jin D, He P, You X, Zhu X, Dai L, He Q, Liu C, Hui N, Sha J, Ni X (2007) Expression of corticotropin-releasing hormone receptor type 1 and type 2 in human pregnant myometrium. Reprod Sci 14:568–577

Jockers R, Maurice P, Boutin JA, Delagrange P (2008) Melatonin receptors, heterodimerization, signal transduction and binding sites: what's new? Br J Pharmacol 154:1182–1195

Jordan B, Devi LA (1998) Molecular mechanisms of opioid receptor signal transduction. Br J Anaesth 81:12–19

Jordan BA, Cvejic S, Devi LA (2000) Opioids and their complicated receptor complexes. Neuropsychopharmacology 23:S5–S18

Jung B, Ahmad N (2006) Melatonin in cancer management: progress and promise. Cancer Res 66:9789–9793

Kallen J, Schlaeppi JM, Bitsch F, Delhon I, Fournier B (2004) Crystal structure of the human RORalpha ligand binding domain in complex with cholesterol sulfate at 2.2 A. J Biol Chem 279:14033–14038

Kamdem LK, Hamilton L, Cheng C, Liu W, Yang W, Johnson JA, Pui CH, Relling MV (2008) Genetic predictors of glucocorticoid-induced hypertension in children with acute lymphoblastic leukemia. Pharmacogenet Genomics 18:507–514

Kanda N, Watanabe S (2003) Histamine enhances the production of nerve growth factor in human keratinocytes. J Invest Dermatol 121:570–577

Kanda N, Watanabe S (2004) Histamine enhances the production of granulocyte-macrophage colony-stimulating factor via protein kinase Calpha and extracellular signal-regulated kinase in human keratinocytes. J Invest Dermatol 122:863–872

Kanda N, Watanabe S (2007) Histamine enhances the production of human beta-defensin-2 in human keratinocytes. Am J Physiol Cell Physiol 293:C1916–C1923

Kaneko K, Travers JB, Matsui MS, Young AR, Norval M, Walker SL (2009) cis-Urocanic acid stimulates primary human keratinocytes independently of serotonin or platelet-activating factor receptors. J Invest Dermatol 129:2567–2573

Karsak M, Gaffal E, Date R, Wang-Eckhardt L, Rehnelt J, Petrosino S, Starowicz K, Steuder R, Schlicker E, Cravatt B, Mechoulam R, Buettner R, Werner S, Di Marzo V, Tuting T, Zimmer A (2007) Attenuation of allergic contact dermatitis through the endocannabinoid system. Science 316:1494–1497

Karst M, Wippermann S, Ahrens J (2010) Role of cannabinoids in the treatment of pain and (painful) spasticity. Drugs 70:2409–2438

Karteris E, Grammatopoulos D, Dai Y, Olah KB, Ghobara TB, Easton A, Hillhouse EW (1998) The human placenta and fetal membranes express the corticotropin-releasing hormone receptor 1alpha (CRH-1alpha) and the CRH-C variant receptor. J Clin Endocrinol Metab 83:1376–1379

Karteris E, Markovic D, Chen J, Hillhouse EW, Grammatopoulos DK (2011) Identification of a novel corticotropin-releasing hormone type 1beta-like receptor variant lacking Exon 13 in human pregnant myometrium regulated by estradiol-17beta and progesterone. Endocrinology 151:4959–4968

Kasumagic-Halilovic E, Prohic A, Begovic B, Ovcina-Kurtovic N (2011) Association between vitiligo and thyroid autoimmunity. J Thyroid Res 2011:938257

Kauser S, Thody AJ, Schallreuter KU, Gummer CL, Tobin DJ (2004) beta-Endorphin as a regulator of human hair follicle melanocyte biology. J Invest Dermatol 123:184–195

Kauser S, Slominski A, Wei ET, Tobin DJ (2006) Modulation of the human hair follicle pigmentary unit by corticotropin-releasing hormone and urocortin peptides. FASEB J 20:882–895

Keen SB, Yelverton CB, Rapp SR, Feldman SR (2008) UV light abuse as a substance-related disorder: clinical implications. Arch Dermatol 144:1047–1048

Khalil Z, Helme RD (1990) Serotonin modulates substance P-induced plasma extravasation and vasodilatation in rat skin by an action through capsaicin-sensitive primary afferent nerves. Brain Res 527:292–298

Khalil EM, De Angelis J, Cole PA (1998) Indoleamine analogs as probes of the substrate selectivity and catalytic mechanism of serotonin N-acetyltransferase. J Biol Chem 273:30321–30327

Kim NH, Lee AY (2010) Histamine effect on melanocyte proliferation and vitiliginous keratinocyte survival. Exp Dermatol 19:1073–1079

Kircik L (2010) A nonsteroidal lamellar matrix cream containing palmitoylethanolamide for the treatment of atopic dermatitis. J Drugs Dermatol 9:334–338

Klein DC (2004) The 2004 Aschoff/Pittendrigh lecture: theory of the origin of the pineal gland—a tale of conflict and resolution. J Biol Rhythms 19:264–279

Klein DC (2007) Arylalkylamine N-acetyltransferase: "the Timezyme". J Biol Chem 282:4233–4237

Ko MC, Husbands SM (2009) Effects of atypical kappa-opioid receptor agonists on intrathecal morphine-induced itch and analgesia in primates. J Pharmacol Exp Ther 328:193–200

Kobayashi H, Kromminga A, Dunlop TW, Tychsen B, Conrad F, Suzuki N, Memezawa A, Bettermann A, Aiba S, Carlberg C, Paus R (2005) A role of melatonin in neuroectodermal-mesodermal interactions: the hair follicle synthesizes melatonin and expresses functional melatonin receptors. FASEB J 19:1710–1712

Koizumi H, Ohkawara A (1999) H2 histamine receptor-mediated increase in intracellular Ca^{2+} in cultured human keratinocytes. J Dermatol Sci 21:127–132

Koizumi H, Shimizu T, Nishino H, Ohkawara A (1998) Cis-urocanic acid attenuates histamine receptor-mediated activation of adenylate cyclase and increase in intracellular Ca^{2+}. Arch Dermatol Res 290:264–269

Kono M, Nagata H, Umemura S, Kawana S, Osamura RY (2001) In situ expression of corticotropin-releasing hormone (CRH) and proopiomelanocortin (POMC) genes in human skin. FASEB J 15:2297–2299

Kopin IJ, Pare CM, Axelrod J, Weissbach H (1961) The fate of melatonin in animals. J Biol Chem 236:3072–3075

Kourosh AS, Harrington CR, Adinoff B (2010) Tanning as a behavioral addiction. Am J Drug Alcohol Abuse 36:284–290

Kraetke O, Wiesner B, Eichhorst J, Furkert J, Bienert M, Beyermann M (2005) Dimerization of corticotropin-releasing factor receptor type 1 is not coupled to ligand binding. J Recept Signal Transduct Res 25:251–276

Kripke ML (1994) Ultraviolet radiation and immunology: something new under the sun—presidential address. Cancer Res 54:6102–6105

Kupczyk P, Reich A, Szepietowski JC (2009) Cannabinoid system in the skin—a possible target for future therapies in dermatology. Exp Dermatol 18:669–679

Kurzen H, Schallreuter KU (2004) Novel aspects in cutaneous biology of acetylcholine synthesis and acetylcholine receptors. Exp Dermatol 13(Suppl 4):27–30

Labrie F, Luu-The V, Labrie C, Pelletier G, El-Alfy M (2000) Intracrinology and the skin. Hormone Res 54:218–229

Labrie F, Luu-The V, Labrie C, Belanger A, Simard J, Lin SX, Pelletier G (2003) Endocrine and intracrine sources of androgens in women: inhibition of breast cancer and other roles of androgens and their precursor dehydroepiandrosterone. Endocr Rev 24:152–182

Lagerstrom MC, Schioth HB (2008) Structural diversity of G protein-coupled receptors and significance for drug discovery. Nat Rev Drug Discov 7:339–357

Lands AM, Arnold A, Mcauliff JP, Luduena FP, Brown TG Jr (1967) Differentiation of receptor systems activated by sympathomimetic amines. Nature 214:597–598

Lassalle MW, Igarashi S, Sasaki M, Wakamatsu K, Ito S, Horikoshi T (2003) Effects of melanogenesis-inducing nitric oxide and histamine on the production of eumelanin and pheomelanin in cultured human melanocytes. Pigment Cell Res 16:81–84

Lee KS, Lee WS, Suh SI, Kim SP, Lee SR, Ryoo YW, Kim BC (2003) Melatonin reduces ultraviolet-B induced cell damages and polyamine levels in human skin fibroblasts in culture. Exp Mol Med 35:263–268

Lefebvre H, Compagnon P, Contesse V, Delarue C, Thuillez C, Vaudry H, Kuhn JM (2001) Production and metabolism of serotonin (5-HT) by the human adrenal cortex: paracrine stimulation of aldosterone secretion by 5-HT. J Clin Endocrinol Metab 86:5001–5007

Legat FJ, Wolf P (2009) Cutaneous sensory nerves: mediators of phototherapeutic effects? Front Biosci 14:4921–4931

Lehmann B, Querings K, Reichrath J (2004) Vitamin D and skin: new aspects for dermatology. Exp Dermatol 13(Suppl 4):11–15

Levins PC, Carr DB, Fisher JE, Momtaz K, Parrish JA (1983) Plasma beta-endorphin and beta-lipoprotein response to ultraviolet radiation. Lancet 2:166

Li W, Chen J, Janjetovic Z, Kim TK, Sweatman T, Lu Y, Zjawiony J, Tuckey RC, Miller D, Slominski A (2010) Chemical synthesis of 20S-hydroxyvitamin D3, which shows antiproliferative activity. Steroids 75:926–935

Ling P, Ngo K, Nguyen S, Thurmond RL, Edwards JP, Karlsson L, Fung-Leung WP (2004) Histamine H4 receptor mediates eosinophil chemotaxis with cell shape change and adhesion molecule upregulation. Br J Pharmacol 142:161–171

Lonne-Rahm SB, Rickberg H, El-Nour H, Marin P, Azmitia EC, Nordlind K (2008) Neuroimmune mechanisms in patients with atopic dermatitis during chronic stress. J Eur Acad Dermatol Venereol 22:11–18

Lu X, Farmer P, Rubin J, Nanes MS (2004) Integration of the NfkappaB p65 subunit into the vitamin D receptor transcriptional complex: identification of p65 domains that inhibit 1,25-dihydroxyvitamin D3-stimulated transcription. J Cell Biochem 92:833–848

Luchetti F, Canonico B, Betti M, Arcangeletti M, Pilolli F, Piroddi M, Canesi L, Papa S, Galli F (2010) Melatonin signaling and cell protection function. FASEB J 24:3603–3624

Luco RF, Allo M, Schor IE, Kornblihtt AR, Misteli T (2011) Epigenetics in alternative pre-mRNA splicing. Cell 144:16–26

Ma X, Idle JR, Krausz KW, Gonzalez FJ (2005) Metabolism of melatonin by human cytochromes p450. Drug Metab Dispos 33:489–494

Ma X, Chen C, Krausz KW, Idle JR, Gonzalez FJ (2008) A metabolomic perspective of melatonin metabolism in the mouse. Endocrinology 149:1869–1879

Maccarrone M, Di Rienzo M, Battista N, Gasperi V, Guerrieri P, Rossi A, Finazzi-Agro A (2003) The endocannabinoid system in human keratinocytes. Evidence that anandamide inhibits epidermal differentiation through CB1 receptor-dependent inhibition of protein kinase C, activation protein-1, and transglutaminase. J Biol Chem 278:33896–33903

Maharaj DS, Anoopkumar-Dukie S, Glass BD, Antunes EM, Lack B, Walker RB, Daya S (2002) The identification of the UV degradants of melatonin and their ability to scavenge free radicals. J Pineal Res 32:257–261

Mammone T, Marenus K, Maes D, Lockshin RA (1998) The induction of terminal differentiation markers by the cAMP pathway in human HaCaT keratinocytes. Skin Pharmacol Appl Skin Physiol 11:152–160

Marcos J, Guo LW, Wilson WK, Porter FD, Shackleton C (2004) The implications of 7-dehydrosterol-7-reductase deficiency (Smith-Lemli-Opitz syndrome) to neurosteroid production. Steroids 69:51–60

Markovic D, Grammatopoulos DK (2009) Focus on the splicing of secretin GPCRs transmembrane-domain 7. Trends Biochem Sci 34:443–452

Markovic D, Lehnert H, Levine MA, Grammatopoulos DK (2008) Structural determinants critical for localization and signaling within the seventh transmembrane domain of the type 1 corticotropin releasing hormone receptor: lessons from the receptor variant R1d. Mol Endocrinol 22:2505–2519

Martinez-Iglesias O, Garcia-Silva S, Tenbaum SP, Regadera J, Larcher F, Paramio JM, Vennstrom B, Aranda A (2009) Thyroid hormone receptor beta1 acts as a potent suppressor of tumor invasiveness and metastasis. Cancer Res 69:501–509

Martin-Ezquerra G, Man MQ, Hupe M, Rodriguez-Martin M, Youm JK, Trullas C, Mackenzie DS, Radek KA, Holleran WM, Elias PM (2011) Psychological stress regulates antimicrobial peptide expression by both glucocorticoid and beta-adrenergic mechanisms. Eur J Dermatol 21 (Suppl 2):48–51

Matsuda M, Imaoka T, Vomachka AJ, Gudelsky GA, Hou Z, Mistry M, Bailey JP, Nieport KM, Walther DJ, Bader M, Horseman ND (2004) Serotonin regulates mammary gland development via an autocrine-paracrine loop. Dev Cell 6:193–203

Maurer M, Opitz M, Henz BM, Paus R (1997) The mast cell products histamine and serotonin stimulate and TNF-alpha inhibits the proliferation of murine epidermal keratinocytes in situ. J Dermatol Sci 16:79–84

Maurice P, Daulat AM, Turecek R, Ivankova-Susankova K, Zamponi F, Kamal M, Clement N, Guillaume JL, Bettler B, Gales C, Delagrange P, Jockers R (2010) Molecular organization and dynamics of the melatonin MT receptor/RGS20/G(i) protein complex reveal asymmetry of receptor dimers for RGS and G(i) coupling. EMBO J 29:3646–3659

Mazurkiewicz JE, Corliss D, Slominski A (2000) Spatiotemporal expression, distribution, and processing of POMC and POMC-derived peptides in murine skin. J Histochem Cytochem 48:905–914

Mazzoni A, Leifer CA, Mullen GE, Kennedy MN, Klinman DM, Segal DM (2003) Cutting edge: histamine inhibits IFN-alpha release from plasmacytoid dendritic cells. J Immunol 170:2269–2273

Memoli S, Napolitano A, D'ischia M, Misuraca G, Palumbo A, Prota G (1997) Diffusible melanin-related metabolites are potent inhibitors of lipid peroxidation. Biochim Biophys Acta 1346:61–68

Mikhailova MV, Mayeux PR, Jurkevich A, Kuenzel WJ, Madison F, Periasamy A, Chen Y, Cornett LE (2007) Heterooligomerization between vasotocin and corticotropin-releasing hormone (CRH) receptors augments CRH-stimulated 3′,5′-cyclic adenosine monophosphate production. Mol Endocrinol 21:2178–2188

Milewich L, Kaimal V, Shaw CB, Sontheimer RD (1986) Epidermal keratinocytes: a source of 5 alpha-dihydrotestosterone production in human skin. J Clin Endocrinol Metab 62:739–746

Milewich L, Shaw CE, Doody KM, Rainey WE, Mason JI, Carr BR (1991) 3 beta-Hydroxysteroid dehydrogenase activity in glandular and extraglandular human fetal tissues. J Clin Endocrinol Metab 73:1134–1140

Missale C, Nash SR, Robinson SW, Jaber M, Caron MG (1998) Dopamine receptors: from structure to function. Physiol Rev 78:189–225

Mockus SM, Vrana KE (1998) Advances in the molecular characterization of tryptophan hydroxylase. J Mol Neurosci 10:163–179

Moniaga CS, Egawa G, Doi H, Miyachi Y, Kabashima K (2011) Histamine modulates the responsiveness of keratinocytes to IL-17 and TNF-alpha through the H1-receptor. J Dermatol Sci 61:79–81

Mossner R, Lesch KP (1998) Role of serotonin in the immune system and in neuroimmune interactions. Brain Behav Immun 12:249–271

Mousa SA, Cheppudira BP, Shaqura M, Fischer O, Hofmann J, Hellweg R, Schafer M (2007) Nerve growth factor governs the enhanced ability of opioids to suppress inflammatory pain. Brain 130:502–513

Musso NR, Brenci S, Indiveri F, Lotti G (1997) L-tyrosine and nicotine induce synthesis of L-Dopa and norepinephrine in human lymphocytes. J Neuroimmunol 74:117–120

Nag K, Sultana N, Kato A, Hirose S (2007) Headless splice variant acting as dominant negative calcitonin receptor. Biochem Biophys Res Commun 362:1037–1043

Nakano K, Yamaoka K, Hanami K, Saito K, Sasaguri Y, Yanagihara N, Tanaka S, Katsuki I, Matsushita S, Tanaka Y (2011) Dopamine induces IL-6-dependent IL-17 production via D1-like receptor on CD4 naive T cells and D1-like receptor antagonist SCH-23390 inhibits cartilage destruction in a human rheumatoid arthritis/SCID mouse chimera model. J Immunol 186:3745–3752

Napolitano A, Memoli S, Nappi AJ, D'ischia M, Prota G (1996) 5-S-cysteinyldopa, a diffusible product of melanocyte activity, is an efficient inhibitor of hydroxylation/oxidation reactions induced by the Fenton system. Biochim Biophys Acta 1291:75–82

Nelson M, Foxwell AR, Tyrer P, Dean RT (2007) Protein-bound 3,4-dihydroxy-phenylanine (DOPA), a redox-active product of protein oxidation, as a trigger for antioxidant defences. Int J Biochem Cell Biol 39:879–889

Nguyen VT, Ndoye A, Grando SA (2000) Novel human alpha9 acetylcholine receptor regulating keratinocyte adhesion is targeted by *Pemphigus vulgaris* autoimmunity. Am J Pathol 157:1377–1391

Nguyen VT, Arredondo J, Chernyavsky AI, Kitajima Y, Grando SA (2003) Keratinocyte acetylcholine receptors regulate cell adhesion. Life Sci 72:2081–2085

Nguyen MN, Slominski A, Li W, Ng YR, Tuckey RC (2009) Metabolism of vitamin D2 to 17,20,24-trihydroxyvitamin D2 by cytochrome p450scc (CYP11A1). Drug Metab Dispos 37:761–767

Nissen JB, Kragballe K (1997) Enkephalins modulate differentiation of normal human keratinocytes in vitro. Exp Dermatol 6:222–229

Nixon AJ, Choy VJ, Parry AL, Pearson AJ (1993) Fiber growth initiation in hair follicles of goats treated with melatonin. J Exp Zool 267:47–56

Nolan BV, Feldman SR (2009) Ultraviolet tanning addiction. Dermatol Clin 27:109–112

Nolan BV, Taylor SL, Liguori A, Feldman SR (2009) Tanning as an addictive behavior: a literature review. Photodermatol Photoimmunol Photomed 25:12–19

Nordlind K, Azmitia EC, Slominski A (2008) The skin as a mirror of the soul: exploring the possible roles of serotonin. Exp Dermatol 17:301–311

Nosjean O, Ferro M, Coge F, Beauverger P, Henlin JM, Lefoulon F, Fauchere JL, Delagrange P, Canet E, Boutin JA (2000) Identification of the melatonin-binding site MT3 as the quinone reductase 2. J Biol Chem 275:31311–31317

Ohnemus U, Uenalan M, Inzunza J, Gustafsson JA, Paus R (2006) The hair follicle as an estrogen target al.nd source. Endocr Rev 27:677–706

Oka S, Wakui J, Ikeda S, Yanagimoto S, Kishimoto S, Gokoh M, Nasui M, Sugiura T (2006) Involvement of the cannabinoid CB2 receptor and its endogenous ligand 2-arachidonoylglycerol in oxazolone-induced contact dermatitis in mice. J Immunol 177:8796–8805

Okayama Y, Church MK (1992) Comparison of the modulatory effect of ketotifen, sodium cromoglycate, procaterol and salbutamol in human skin, lung and tonsil mast cells. Int Arch Allergy Immunol 97:216–225

Owens MJ, Nemeroff CB (1991) Physiology and pharmacology of corticotropin-releasing factor. Pharmacol Rev 43:425–473

Pacher P, Mechoulam R (2011) Is lipid signaling through cannabinoid 2 receptors part of a protective system? Prog Lipid Res 50:193–211

Page-Mccaw PS, Chung SC, Muto A, Roeser T, Staub W, Finger-Baier KC, Korenbrot JI, Baier H (2004) Retinal network adaptation to bright light requires tyrosinase. Nat Neurosci 7:1329–1336

Pandi-Perumal SR, Srinivasan V, Maestroni GJ, Cardinali DP, Poeggeler B, Hardeland R (2006) Melatonin: nature's most versatile biological signal? FEBS J 273:2813–2838

Paradisi A, Pasquariello N, Barcaroli D, Maccarrone M (2008) Anandamide regulates keratinocyte differentiation by inducing DNA methylation in a CB1 receptor-dependent manner. J Biol Chem 283:6005–6012

Park HY, Kosmadaki M, Yaar M, Gilchrest BA (2009) Cellular mechanisms regulating human melanogenesis. Cell Mol Life Sci 66:1493–1506

Patel T, Yosipovitch G (2010) Therapy of pruritus. Expert Opin Pharmacother 11:1673–1682

Pattwell DM, Lynaugh KJ, Watson RE, Paus R (2010) HaCaT keratinocytes express functional receptors for thyroid-stimulating hormone. J Dermatol Sci 59:52–55

Payne AH, Hales DB (2004) Overview of steroidogenic enzymes in the pathway from cholesterol to active steroid hormones. Endocr Rev 25:947–970

Perrin MH, Vale WW (1999) Corticotropin releasing factor receptors and their ligand family. Ann NY Acad Sci 885:312–328

Perrin MH, Digruccio MR, Koerber SC, Rivier JE, Kunitake KS, Bain DL, Fischer WH, Vale WW (2003) A soluble form of the first extracellular domain of mouse type 2beta corticotropin-releasing factor receptor reveals differential ligand specificity. J Biol Chem 278:15595–15600

Pinczewski J, Slominski A (2010) The potential role of vitamin D in the progression of benign and malignant melanocytic neoplasms. Exp Dermatol 19:860–864

Pisarchik A, Slominski AT (2001) Alternative splicing of CRH-R1 receptors in human and mouse skin: identification of new variants and their differential expression. FASEB J 15:2754–2756

Pisarchik A, Slominski A (2002) Corticotropin releasing factor receptor type 1: molecular cloning and investigation of alternative splicing in the hamster skin. J Invest Dermatol 118:1065–1072

Pisarchik A, Slominski A (2004) Molecular and functional characterization of novel CRFR1 isoforms from the skin. Eur J Biochem 271:2821–2830

Pisarchik A, Wortsman J, Slominski A (2004) A novel microarray to evaluate stress-related genes in skin: effect of ultraviolet light radiation. Gene 341:199–207

Poeggeler B, Knuever J, Gaspar E, Biro T, Klinger M, Bodo E, Wiesner RJ, Wenzel BE, Paus R (2010) Thyrotropin powers human mitochondria. FASEB J 24:1525–1531

Polakiewicz RD, Behar OZ, Comb MJ, Rosen H (1992) Regulation of proenkephalin expression in cultured skin mesenchymal cells. Mol Endocrinol 6:399–408

Polymeropoulos MH, Torres R, Yanovski JA, Chandrasekharappa SC, Ledbetter DH (1995) The human corticotropin-releasing factor receptor (CRHR) gene maps to chromosome 17q12-q22. Genomics 28:123–124

Pozo D, Garcia-Maurino S, Guerrero JM, Calvo JR (2004) mRNA expression of nuclear receptor RZR/RORalpha, melatonin membrane receptor MT, and hydroxindole-O-methyltransferase in different populations of human immune cells. J Pineal Res 37:48–54

Pritchard LE, White A (2007) Neuropeptide processing and its impact on melanocortin pathways. Endocrinology 148:4201–4207

Prokopenko I, Langenberg C, Florez JC, Saxena R, Soranzo N, Thorleifsson G et al (2009) Variants in MTNR1B influence fasting glucose levels. Nat Genet 41:77–81

Przewlocki R (2004) Opioid abuse and brain gene expression. Eur J Pharmacol 500:331–349

Przewlocki R, Przewlocka B (2005) Opioids in neuropathic pain. Curr Pharm Des 11:3013–3025

Pucci M, Pirazzi V, Pasquariello N, Maccarrone M (2011) Endocannabinoid signaling and epidermal differentiation. Eur J Dermatol Suppl 2:29–34

Pullar CE, Isseroff RR (2006) The beta 2-adrenergic receptor activates pro-migratory and pro-proliferative pathways in dermal fibroblasts via divergent mechanisms. J Cell Sci 119:592–602

Pullar CE, Isseroff RR, Nuccitelli R (2001) Cyclic AMP-dependent protein kinase A plays a role in the directed migration of human keratinocytes in a DC electric field. Cell Motil Cytoskeleton 50:207–217

Pullar CE, Manabat-Hidalgo CG, Bolaji RS, Isseroff RR (2008) beta-Adrenergic receptor modulation of wound repair. Pharmacol Res 58:158–164

Quevedo ME, Slominski A, Pinto W, Wei E, Wortsman J (2001) Pleiotropic effects of corticotropin releasing hormone on normal human skin keratinocytes. In Vitro Cell Dev Biol Anim 37:50–54

Rahn EJ, Hohmann AG (2009) Cannabinoids as pharmacotherapies for neuropathic pain: from the bench to the bedside. Neurotherapeutics 6:713–737

Randall VA, Thornton MJ, Messenger AG, Hibberts NA, Loudon AS, Brinklow BR (1993) Hormones and hair growth: variations in androgen receptor content of dermal papilla cells cultured from human and red deer (*Cervus elaphus*) hair follicles. J Invest Dermatol 101:114S–120S

Rasul A, El-Nour H, Blakely RD, Lonne-Rahm SB, Forsberg J, Johansson B, Theodorsson E, Nordlind K (2011) Effect of chronic mild stress on serotonergic markers in the skin and brain of the NC/Nga atopic-like mouse strain. Arch Dermatol Res 303:625–633

Reich A, Wojcik-Maciejewicz A, Slominski AT (2010) Stress and the skin. G Ital Dermatol Venereol 145:213–219

Reichrath J (2007) Vitamin D and the skin: an ancient friend, revisited. Exp Dermatol 16:618–625

Reiter RJ (1991) Pineal melatonin: cell biology of its synthesis and of its physiological interactions. Endocr Rev 12:151–180

Reiter RJ, Tan DX, Pilar Terron M, Flores LJ, Czarnocki Z (2007) Melatonin and its metabolites: new findings regarding their production and their radical scavenging actions. Acta Biochim Pol 54:1–9

Ricketts ML, Verhaeg JM, Bujalska I, Howie AJ, Rainey WE, Stewart PM (1998) Immunohistochemical localization of type 1 11beta-hydroxysteroid dehydrogenase in human tissues. J Clin Endocrinol Met 83:1325–1335

Rogoff D, Gomez-Sanchez CE, Foecking MF, Wortsman J, Slominski A (2001) Steroidogenesis in the human skin: 21-hydroxylation in cultured keratinocytes. J Steroid Biochem Mol Biol 78:77–81

Roloff B, Fechner K, Slominski A, Furkert J, Botchkarev VA, Bulfone-Paus S, Zipper J, Krause E, Paus R (1998) Hair cycle-dependent expression of corticotropin-releasing factor (CRF) and CRF receptors in murine skin. FASEB J 12:287–297

Roosterman D, Goerge T, Schneider SW, Bunnett NW, Steinhoff M (2006) Neuronal control of skin function: the skin as a neuroimmunoendocrine organ. Physiol Rev 86:1309–1379

Roseboom PH, Namboodiri MA, Zimonjic DB, Popescu NC, Rodriguez IR, Gastel JA, Klein DC (1998) Natural melatonin 'knockdown' in C57BL/6J mice: rare mechanism truncates serotonin *N*-acetyltransferase. Brain Res Mol Brain Res 63:189–197

Ross PC, Kostas CM, Ramabhadran TV (1994) A variant of the human corticotropin-releasing factor (CRF) receptor: cloning, expression and pharmacology. Biochem Biophys Res Commun 205:1836–1842

Rousseau K, Kauser S, Pritchard LE, Warhurst A, Oliver RL, Slominski A, Wei ET, Thody AJ, Tobin DJ, White A (2007) Proopiomelanocortin (POMC), the ACTH/melanocortin precursor, is secreted by human epidermal keratinocytes and melanocytes and stimulates melanogenesis. FASEB J 21:1844–1856

Ryoo YW, Suh SI, Mun KC, Kim BC, Lee KS (2001) The effects of the melatonin on ultraviolet-B irradiated cultured dermal fibroblasts. J Dermatol Sci 27:162–169

Safer JD, Persons K, Holick MF (2009) A thyroid hormone deiodinase inhibitor can decrease cutaneous cell proliferation in vitro. Thyroid 19:181–185

Sagar DR, Staniaszek LE, Okine BN, Woodhams S, Norris LM, Pearson RG, Garle MJ, Alexander SP, Bennett AJ, Barrett DA, Kendall DA, Scammell BE, Chapman V (2010) Tonic modulation of spinal hyperexcitability by the endocannabinoid receptor system in a rat model of osteoarthritis pain. Arthritis Rheum 62:3666–3676

Saha B, Mondal AC, Basu S, Dasgupta PS (2001) Circulating dopamine level, in lung carcinoma patients, inhibits proliferation and cytotoxicity of CD4+ and CD8+ T cells by D1 dopamine receptors: an in vitro analysis. Int Immunopharmacol 1:1363–1374

Salemi S, Aeschlimann A, Reisch N, Jungel A, Gay RE, Heppner FL, Michel BA, Gay S, Sprott H (2005) Detection of kappa and delta opioid receptors in skin–outside the nervous system. Biochem Biophys Res Commun 338:1012–1017

Salim S, Ali SA (2011) Vertebrate melanophores as potential model for drug discovery and development: a review. Cell Mol Biol Lett 16:162–200

Sanchez AJ, Garcia-Merino A (2012) Neuroprotective agents: Cannabinoids. Clin Immunol 142:57–67

Scarparo AC, Visconti MA, De Oliveira AR, Castrucci AM (2000) Adrenoceptors in normal and malignant human melanocytes. Arch Dermatol Res 292:265–267

Schallreuter KU, Wood JM, Lemke R, Lepoole C, Das P, Westerhof W, Pittelkow MR, Thody AJ (1992) Production of catecholamines in the human epidermis. Biochem Biophys Res Comm 189:72–78

Schallreuter KU, Wood JM, Pittelkow MR, Swanson NN, Steinkraus V (1993) Increased in vitro expression of beta 2-adrenoceptors in differentiating lesional keratinocytes of vitiligo patients. Arch Dermatol Res 285:216–220

Schallreuter KU, Wood JM, Pittelkow MR, Gutlich M, Lemke KR, Rodl W, Swanson NN, Hitzemann K, Ziegler I (1994) Regulation of melanin biosynthesis in the human epidermis by tetrahydrobiopterin. Science 263:1444–1446

Schallreuter KU, Lemke KR, Pittelkow MR, Wood JM, Korner C, Malik R (1995) Catecholamines in human keratinocyte differentiation. J Invest Dermatol 104:953–957

Schallreuter KU, Korner C, Pittelkow MR, Swanson NN, Gardner ML (1996) The induction of the alpha-1-adrenoceptor signal transduction system on human melanocytes. Exp Dermatol 5:20–23

Schallreuter KU, Pittelkow MR, Swanson NN, Beazley WD, Korner C, Ehrke C, Buttner G (1997) Altered catecholamine synthesis and degradation in the epidermis of patients with atopic eczema. Arch Dermatol Res 289:663–666

Schallreuter KU, Beazley WD, Hibberts NA, Tobin DJ, Paus R, Wood JM (1998) Pterins in human hair follicle cells and in the synchronized murine hair cycle. J Invest Dermatol 111:545–550

Schallreuter KU, Bahadoran P, Picardo M, Slominski A, Elassiuty YE, Kemp EH, Giachino C, Liu JB, Luiten RM, Lambe T, Le Poole IC, Dammak I, Onay H, Zmijewski MA, Dell'anna ML, Zeegers MP, Cornall RJ, Paus R, Ortonne JP, Westerhof W (2008a) Vitiligo pathogenesis: autoimmune disease, genetic defect, excessive reactive oxygen species, calcium imbalance, or what else? Exp Dermatol 17:139–140, discussion 141-160

Schallreuter KU, Kothari S, Chavan B, Spencer JD (2008b) Regulation of melanogenesis–controversies and new concepts. Exp Dermatol 17:395–404

Schmelz M (2010) Itch and pain. Neurosci Biobehav Rev 34:171–176

Schmelz M, Paus R (2007) Opioids and the skin: "itchy" perspectives beyond analgesia and abuse. J Invest Dermatol 127:1287–1289

Schmid B, Blomeyer D, Treutlein J, Zimmermann US, Buchmann AF, Schmidt MH, Esser G, Rietschel M, Banaschewski T, Schumann G, Laucht M (2009) Interacting effects of CRHR1 gene and stressful life events on drinking initiation and progression among 19-year-olds. Int J Neuropsychopharmacol 3:1–12

Scislowski PW, Slominski A, Bomirski A (1984) Biochemical characterization of three hamster melanoma variants—II. Glycolysis and oxygen consumption. Int J Biochem 16:327–331

Scislowski PW, Slominski A, Bomirski A, Zydowo M (1985) Metabolic characterization of three hamster melanoma variants. Neoplasma 32:593–598

Seck T, Pellegrini M, Florea AM, Grignoux V, Baron R, Mierke DF, Horne WC (2005) The delta e13 isoform of the calcitonin receptor forms a six-transmembrane domain receptor with dominant-negative effects on receptor surface expression and signaling. Mol Endocrinol 19:2132–2144

Seever K, Hardeland R (2008) Novel pathway for N(1)-acetyl-5-methoxykynuramine: UVB-induced liberation of carbon monoxide from precursor N(1)-acetyl-N(2)-formyl-5-methoxyky-nuramine. J Pineal Res 44:450–455

Seiffert K, Hosoi J, Torii H, Ozawa H, Ding W, Campton K, Wagner JA, Granstein RD (2002) Catecholamines inhibit the antigen-presenting capability of epidermal Langerhans cells. J Immunol 168:6128–6135

Selye H (1936) A syndrome produced by various noxious agents. Nat London 138:32–33

Semak I, Korik E, Naumova M, Wortsman J, Slominski A (2004) Serotonin metabolism in rat skin: characterization by liquid chromatography-mass spectrometry. Arch Biochem Biophys 421:61–66

Semak I, Naumova M, Korik E, Terekhovich V, Wortsman J, Slominski A (2005) A novel metabolic pathway of melatonin: oxidation by cytochrome C. Biochemistry 44:9300–9307

Semak I, Korik E, Antonova M, Wortsman J, Slominski A (2008) Metabolism of melatonin by cytochrome P450s in rat liver mitochondria and microsomes. J Pineal Res 45:515–523

Seuwen K, Pouysségur J (1990) Serotonin as a growth factor. Biochem Pharmacol 39(6):985–990

Seyle H (1976) The stress of life. McGraw-Hill Book Company, New York, NY

Shackleton CH, Roitman E, Kelley R (1999) Neonatal urinary steroids in Smith-Lemli-Opitz syndrome associated with 7-dehydrocholesterol reductase deficiency. Steroids 64:481–490

Shackleton C, Roitman E, Guo LW, Wilson WK, Porter FD (2002) Identification of 7(8) and 8(9) unsaturated adrenal steroid metabolites produced by patients with 7-dehydrosterol-delta7-reductase deficiency (Smith-Lemli-Opitz syndrome). J Steroid Bioche Mol Biol 82:225–232

Sharpley CF, Kauter KG, Mcfarlane JR (2009) An initial exploration of in vivo hair cortisol responses to a brief pain stressor: latency, localization and independence effects. Physiol Res 58:757–761

Shinoda S, Kameyoshi Y, Hide M, Morita E, Yamamoto S (1998) Histamine enhances UVB-induced IL-6 production by human keratinocytes. Arch Dermatol Res 290:429–434

Sidler D, Renzulli P, Schnoz C, Berger B, Schneider-Jakob S, Fluck C, Inderbitzin D, Corazza N, Candinas D, Brunner T (2011) Colon cancer cells produce immunoregulatory glucocorticoids. Oncogene 30:2411–2419

Siemionow M, Gharb BB, Rampazzo A (2011) The face as a sensory organ. Plast Reconstr Surg 127:652–662

Sikand P, Dong X, Lamotte RH (2011) BAM8-22 peptide produces itch and nociceptive sensations in humans independent of histamine release. J Neurosci 31:7563–7567

Simard J, Couet J, Durocher F, Labrie Y, Sanchez R, Breton N, Turgeon C, Labrie F (1993) Structure and tissue-specific expression of a novel member of the rat 3 beta-hydroxysteroid dehydrogenase/delta 5-delta 4 isomerase (3 beta-HSD) family. The exclusive 3 beta-HSD gene expression in the skin. J Biol Chem 268:19659–19668

Simard J, Ricketts ML, Gingras S, Soucy P, Feltus FA, Melner MH (2005) Molecular biology of the 3beta-hydroxysteroid dehydrogenase/delta5-delta4 isomerase gene family. Endocrine Rev 26:525–582

Sivamani RK, Lam ST, Isseroff RR (2007) Beta adrenergic receptors in keratinocytes. Dermatol Clin 25:643–653

Sivamani RK, Porter SM, Isseroff RR (2009) An epinephrine-dependent mechanism for the control of UV-induced pigmentation. J Invest Dermatol 129:784–787

Skobowiat C, Dowdy JC, Sayre RM, Tuckey RC, Slominski AT (2011) Cutaneous hypothalamic pituitary adrenal (HPA) axis homologue—regulation by ultraviolet radiation. Am J Physiol Endocrinol Metab 301:E484–E493

Slominski A (1991) POMC gene expression in mouse and hamster melanoma cells. FEBS Lett 291:165–168

Slominski A (2003) Beta-endorphin/mu-opiate receptor system in the skin. J Invest Dermatol. 120: xii–xiii

Slominski A (2005) Neuroendocrine system of the skin. Dermatology 211:199–208

Slominski A (2007) A nervous breakdown in the skin: stress and the epidermal barrier. J Clin Invest 117:3166–3169

Slominski A (2009a) Are suberythemal doses of ultraviolet B good for your skin? Pigment Cell Melanoma Res 22:154–155

Slominski A (2009b) On the role of the corticotropin-releasing hormone signalling system in the aetiology of inflammatory skin disorders. Br J Dermatol 160:229–232

Slominski A (2009c) Neuroendocrine activity of the melanocyte. Exp Dermatol 18:760–763

Slominski A, Friedrich T (1992) L-DOPA inhibits in vitro phosphorylation of melanoma glycoproteins. Pigment Cell Res 5:396–400

Slominski A, Goodman-Snitkoff GG (1992) Dopa inhibits induced proliferative activity of murine and human lymphocytes. Anticancer Res 12:753–756

Slominski A, Paus R (1990) Are L-tyrosine and L-dopa hormone-like bioregulators. J Theor Biol 143:123–138

Slominski A, Paus R (1994) Towards defining receptors for L-tyrosine and L-dopa. Mol Cell Endocrinol 99:C7–C11

Slominski A, Pawelek J (1998) Animals under the sun: effects of ultraviolet radiation on mammalian skin. Clin Dermatol 16:503–515

Slominski A, Pruski D (1993) Melatonin inhibits proliferation and melanogenesis in rodent melanoma cells. Exp Cell Res 206:189–294

Slominski A, Wortsman J (2000) Neuroendocrinology of the skin. Endocrine Rev 21:457–487

Slominski A, Paus R, Bomirski A (1989) Hypothesis: possible role for the melatonin receptor in vitiligo: discussion paper. J R Soc Med 82:539–541

Slominski A, Paus R, Mazurkiewicz J (1992) Proopiomelanocortin expression in the skin during induced hair growth in mice. Experientia 48:50–54

Slominski A, Paus R, Schadendorf D (1993a) Melanocytes as "sensory" and regulatory cells in the epidermis. J Theor Biol 164:103–120

Slominski A, Paus R, Wortsman J (1993b) On the potential role of proopiomelanocortin in skin physiology and pathology. Mol Cell Endocrinol 93:C1–C6

Slominski A, Wortsman J, Mazurkiewicz JE, Matsuoka L, Dietrich J, Lawrence K, Gorbani A, Paus R (1993c) Detection of proopiomelanocortin-derived antigens in normal and pathologic human skin. J Lab Clin Med 122:658–666

Slominski A, Chassalevris N, Mazurkiewicz J, Maurer M, Paus R (1994) Murine skin as a target for melatonin bioregulation. Exp Dermatol 3:45–50

Slominski A, Ermak G, Hwang J, Chakraborty A, Mazurkiewicz J, Mihm M (1995) Proopiomelanocortin, corticotropin releasing hormone and corticotropin releasing hormone receptor genes are expressed in human skin. FEBS Lett 374:113–116

Slominski A, Baker J, Ermak G, Chakraborty A, Pawelek J (1996a) Ultraviolet B stimulates production of corticotropin releasing factor (CRF) by human melanocytes. FEBS Lett 399:175–176

Slominski A, Baker J, Rosano TG, Guisti LW, Ermak G, Grande M, Gaudet SJ (1996b) Metabolism of serotonin to N-acetylserotonin, melatonin, and 5-methoxytryptamine in hamster skin culture. J Biol Chem 271:12281–12286

Slominski A, Ermak G, Hwang J, Mazurkiewicz J, Corliss D, Eastman A (1996c) The expression of proopiomelanocortin (POMC) and of corticotropin releasing hormone receptor (CRH-R) genes in mouse skin. Biochim Biophys Acta 1289:247–251

Slominski A, Ermak G, Mihm MC (1996d) ACTH receptor, CYP11A1, CYP17 and CYP21A2 genes are expressed in skin. J Clin Endocrinol Metol 81:2746–2749

Slominski A, Botchkareva NV, Botchkarev VA, Chakraborty A, Luger T, Uenalan M, Paus R (1998a) Hair cycle-dependent production of ACTH in mouse skin. Biochim Biophys Acta 1448:147–152

Slominski A, Ermak G, Mazurkiewicz JE, Baker J, Wortsman J (1998b) Characterization of corticotropin-releasing hormone (CRH) in human skin. J Clin Endocrinol Metol 83:1020–1024

Slominski A, Paus R, Mihm MC (1998c) Inhibition of melanogenesis as an adjuvant strategy in the treatment of melanotic melanomas: selective review and hypothesis. Anticancer Res 18:3709–3715

Slominski A, Gomez-Sanchez C, Foecking MF, Wortsman J (1999a) Metabolism of progesterone to DOC, corticosterone and 18OHDOC in cultured human melanoma cells. FEBS Lett 445:364–366

Slominski A, Gomez-Sanchez CE, Foecking MF, Wortsman J (1999b) Metabolism of progesterone to DOC, corticosterone and 18OHDOC in cultured human melanoma cells. FEBS Lett 455:364–366

Slominski AT, Botchkarev V, Choudhry M, Fazal N, Fechner K, Furkert J, Krause E, Roloff B, Sayeed M, Wei E, Zbytek B, Zipper J, Wortsman J, Paus R (1999c) Cutaneous expression of CRH and CRH-R. Is there a "skin stress response system?". Ann N Y Acad Sci 885:287–311

Slominski A, Gomez-Sanchez C, Foecking MF, Wortsman J (2000a) Active steroidogenesis in the normal rat skin. Biochim Biophys Acta 1474:1

Slominski A, Roloff B, Curry J, Dahiya M, Szczesniewski A, Wortsman J (2000b) The skin produces urocortin. J Clin Endocrinol Metab 85:815–823

Slominski A, Wortsman J, Luger T, Paus R, Solomon S (2000c) Corticotropin releasing hormone and proopiomelanocortin involvement in the cutaneous response to stress. Physiol Rev 80:979–1020

Slominski AT, Roloff B, Zbytek B, Wei ET, Fechner K, Curry J, Wortsman J (2000d) Corticotropin releasing hormone and related peptides can act as bioregulatory factors in human keratinocytes. In Vitro Cell Dev Biol Anim 36:211–216

Slominski A, Wortsman J, Pisarchik A, Zbytek B, Linton EA, Mazurkiewicz JE, Wei ET (2001) Cutaneous expression of corticotropin-releasing hormone (CRH), urocortin, and CRH receptors. FASEB J 15:1678–1693

Slominski A, Pisarchik A, Semak I, Sweatman T, Szczesniewski A, Wortsman J (2002a) Serotoninergic system in hamster skin. J Invest Dermatol 119:934–942

Slominski A, Pisarchik A, Semak I, Sweatman T, Worstman J, Szczesniewski A, Slugocki G, Mcnulty J, Kauser S, Tobin DJ (2002b) Serotoninergic and melatoninergic systems are fully expressed in human skin. FASEB J 16:896–898

Slominski A, Semak I, Pisarchik A, Sweatman T, Szczesniewski A, Wortsman J (2002c) Conversion of L-tryptophan to serotonin and melatonin in human melanoma cells. FEBS Lett 511:102–106

Slominski A, Worstman J, Foecking MF, Shackleton CH, Gomez-Sanchez CE, Szczesniewski A (2002d) Gas chromatography/mass spectrometry characterization of corticosteroid metabolism in human immortalized keratinocytes. J Invest Dermatol 118:310–315

Slominski A, Wortsman J, Kohn L, Ain KB, Venkataraman GM, Pisarchik A, Chung JH, Giuliani C, Thornton M, Slugocki G, Tobin DJ (2002e) Expression of hypothalamic-pituitary-thyroid axis related genes in the human skin. J Invest Dermatol 119:1449–1455

Slominski A, Pisarchik A, Johansson O, Jing C, Semak I, Slugocki G, Wortsman J (2003a) Tryptophan hydroxylase expression in human skin cells. Biochim Biophys Acta 1639:80–86

Slominski A, Wortsman J, Linton E, Pisarchik A, Zbytek B (2003b) The skin as a model for the immunomodulatory effects of corticotropin-releasing hormone. In: Schaefer M, Stein C (eds) Mind over matter—regulation of peripheral inflammation by the CNS. Birkhaeuser Verlag, Basel, Boston, Berlin

Slominski A, Pisarchik A, Semak I, Sweatman T, Wortsman J (2003c) Characterization of the serotoninergic system in the C57BL/6 mouse skin. Eur J Biochem 270:3335–3344

Slominski A, Pisarchik A, Zbytek B, Tobin DJ, Kauser S, Wortsman J (2003d) Functional activity of serotoninergic and melatoninergic systems expressed in the skin. J Cell Physiol 196:144–153

Slominski A, Pisarchik A, Tobin DJ, Mazurkiewicz JE, Wortsman J (2004a) Differential expression of a cutaneous corticotropin-releasing hormone system. Endocrinology 145:941–950

Slominski A, Pisarchik A, Wortsman J (2004b) Expression of genes coding melatonin and serotonin receptors in rodent skin. Biochim Biophys Acta 1680:67–70

Slominski A, Tobin DJ, Shibahara S, Wortsman J (2004c) Melanin pigmentation in mammalian skin and its hormonal regulation. Physiol Rev 84:1155–1228

Slominski A, Zjawiony J, Wortsman J, Semak I, Stewart J, Pisarchik A, Sweatman T, Marcos J, Dunbar C, Tuckey RC (2004d) A novel pathway for sequential transformation of 7-dehydrocholesterol and expression of the P450scc system in mammalian skin. Eur J Biochem 271:4178–4188

Slominski A, Fischer TW, Zmijewski MA, Wortsman J, Semak I, Zbytek B, Slominski RM, Tobin DJ (2005a) On the role of melatonin in skin physiology and pathology. Endocrine 27:137–148

Slominski A, Plonka PM, Pisarchik A, Smart JL, Tolle V, Wortsman J, Low MJ (2005b) Preservation of eumelanin hair pigmentation in proopiomelanocortin-deficient mice on a nonagouti (a/a) genetic background. Endocrinology 146:1245–1253

Slominski A, Wortsman J, Tobin DJ (2005c) The cutaneous serotoninergic/melatoninergic system: securing a place under the sun. FASEB J 19:176–194

Slominski A, Zbytek B, Semak I, Sweatman T, Wortsman J (2005d) CRH stimulates POMC activity and corticosterone production in dermal fibroblasts. J Neuroimmunol 162:97–102

Slominski A, Zbytek B, Szczesniewski A, Semak I, Kaminski J, Sweatman T, Wortsman J (2005e) CRH stimulation of corticosteroids production in melanocytes is mediated by ACTH. Am J Physiol Endocrinol Metab 288:E701–E706

Slominski A, Semak I, Zjawiony J, Wortsman J, Li W, Szczesniewski A, Tuckey RT (2005f) The cytochrome P450scc system opens an alternate pathway of vitamin D3metabolism. FEBS J 272:480–4090

Slominski A, Semak I, Zjawiony J, Wortsman J, Gandy MN, Li J, Zbytek B, Li W, Tuckey RT (2005g) Enzymatic metabolism of ergosterol by cytochrome P450scc (CYP11A1) to biologically active 17α,24-dihydroxyergosterol. Chem Biol 12:931–939

Slominski A, Zbytek B, Pisarchik A, Slominski RM, Zmijewski MA, Wortsman J (2006a) CRH functions as a growth factor/cytokine in the skin. J Cell Physiol 206:780–791

Slominski A, Zbytek B, Szczesniewski A, Wortsman J (2006b) Cultured human dermal fibroblasts do produce cortisol. J Invest Dermatol 126:1177–1178

Slominski A, Zbytek B, Zmijewski M, Slominski RM, Kauser S, Wortsman J, Tobin DJ (2006c) Corticotropin releasing hormone and the skin. Front Biosci 11:2230–2248

Slominski A, Semak I, Wortsman J, Zjawiony J, Li W, Zbytek B, Tuckey T (2006d) An alternate pathway of vitamin D2 metabolism: Cytochrome P450scc (CYP11A1) mediated conversion to20-hydroxyvitamin D2 and 17,20-dihydroxyvitamin D2. FEBS J 273:2891–2901

Slominski A, Wortsman J, Tuckey RC, Paus R (2007a) Differential expression of HPA axis homolog in the skin. Mol Cellular Endocrinol 265–266:143–149

Slominski AT, Zmijewski MA, Pisarchik A, Wortsman J (2007b) Molecular cloning and initial characterization of African green monkey (Cercopithecus aethiops) corticotropin releasing factor receptor type 1 (CRFI) from COS-7 cells. Gene 389:154–162

Slominski A, Tobin DJ, Zmijewski MA, Wortsman J, Paus R (2008a) Melatonin in the skin: synthesis, metabolism and functions. Trends Endo Metab 19:17–24

Slominski A, Wortsman J, Paus R, Elias PM, Tobin DJ, Feingold KR (2008b) Skin as an endocrine organ: implications for its function. Drug Discov Today Dis Mech 5:137–144

Slominski A, Tuckey RC, Zmijewski MA, Li W, Zjawiony J, Janjetovic Z, Zbytek B, Nguyen MN, Miller D, Chen J (2009a) Enzymatic production or chemical synthesis and uses for 5,7-dienes and UVB conversion products thereof. In PCT/US2009/001324

Slominski A, Zbytek B, Slominski R (2009b) Inhibitors of melanogenesis increase toxicity of cyclophosphamide and lymphocytes against melanoma cells. Int J Cancer 124:1470–1477

Slominski AT, Zmijewski MA, Semak I, Sweatman T, Janjetovic Z, Li W, Zjawiony JK, Tuckey RC (2009c) Sequential metabolism of 7-dehydrocholesterol to steroidal 5,7-dienes in adrenal glands and its biological implication in the skin. PLoS One 4:e4309

Slominski AT, Janjetovic Z, Fuller BE, Zmijewski MA, Tuckey RC, Nguyen MN, Sweatman T, Li W, Zjawiony J, Miller D, Chen TC, Lozanski G, Holick MF (2010) Products of vitamin D3 or 7-dehydrocholesterol metabolism by cytochrome P450scc show anti-leukemia effects, having low or absent calcemic activity. PLoS One 5:e9907

Slominski A, Zmijewski M, Pawelek J (2011a) L-tyrosine and L-DOPA as hormone-like regulators of melanocytes functions. Pigment Cell Melanoma Res. doi:10.1111/j.1755-148X.2011.00898.x

Slominski AT, Kim TK, Janjetovic Z, Tuckey RC, Bieniek R, Yue J, Li W, Chen J, Nguyen MN, Tang EK, Miller D, Chen TC, Holick M (2011b) 20-Hydroxyvitamin D2 is a noncalcemic analog of vitamin D with potent antiproliferative and prodifferentiation activities in normal and malignant cells. Am J Physiol Cell Physiol 300:C526–C541

Slominski AT, Zmijewski MA, Zbytek B, Brozyna AA, Granese J, Pisarchik A, Szczesniewski A, Tobin DJ (2011c) Regulated proenkephalin expression in human skin and cultured skin cells. J Invest Dermatol 131:613–622

Slominski AT, Li W, Bhattacharya SK, Smith RA, Johnson PL, Chen J, Nelson KE, Tuckey RC, Miller D, Jiao Y, Gu W, Postlethwaite A (2011d) Vitamin D analogs 17,20S(OH)2pD and 17,20R(OH)2pD are noncalcemic and exhibit antifibrotic activity. J Invest Dermatol 131:1167–1169

Smith AI, Funder JW (1988) Proopiomelanocortin processing in the pituitary, central nervous system, and peripheral tissues. Endocr Rev 9:159–179

Spiess J, Rivier J, Rivier C, Vale W (1981) Primary structure of corticotropin-releasing factor from ovine hypothalamus. Proc Natl Acad Sci USA 78:6517–6521

Sreevidya CS, Khaskhely NM, Fukunaga A, Khaskina P, Ullrich SE (2008) Inhibition of photocarcinogenesis by platelet-activating factor or serotonin receptor antagonists. Cancer Res 68:3978–3984

Sreevidya CS, Fukunaga A, Khaskhely NM, Masaki T, Ono R, Nishigori C, Ullrich SE (2010) Agents that reverse UV-Induced immune suppression and photocarcinogenesis affect DNA repair. J Invest Dermatol 130:1428–1437

Srivastava BK, Soni R, Patel JZ, Joharapurkar A, Sadhwani N, Kshirsagar S, Mishra B, Takale V, Gupta S, Pandya P, Kapadnis P, Solanki M, Patel H, Mitra P, Jain MR, Patel PR (2009) Hair growth stimulator property of thienyl substituted pyrazole carboxamide derivatives as a CB1 receptor antagonist with in vivo antiobesity effect. Bioorg Med Chem Lett 19:2546–2550

Stander S, Gunzer M, Metze D, Luger T, Steinhoff M (2002) Localization of mu-opioid receptor 1A on sensory nerve fibers in human skin. Regul Pept 110:75–83

Stander S, Schmelz M, Metze D, Luger T, Rukwied R (2005) Distribution of cannabinoid receptor 1 (CB1) and 2 (CB2) on sensory nerve fibers and adnexal structures in human skin. J Dermatol Sci 38:177–188

Steinhoff M, Bienenstock J, Schmelz M, Maurer M, Wei E, Biro T (2006) Neurophysiological, neuroimmunological, and neuroendocrine basis of pruritus. J Invest Dermatol 126:1705–1718

Steinkraus V, Steinfath M, Korner C, Mensing H (1992) Binding of beta-adrenergic receptors in human skin. J Invest Dermatol 98:475–480

Steinkraus V, Mak JC, Pichlmeier U, Mensing H, Ring J, Barnes PJ (1996) Autoradiographic mapping of beta-adrenoceptors in human skin. Arch Dermatol Res 288:549–553

Stenn KS, Paus R (2001) Controls of hair follicle cycling. Physiol Rev 81:449–494

Su TF, Zhang LH, Peng M, Wu CH, Pan W, Tian B, Shi J, Pan HL, Li M (2011) Cannabinoid CB2 receptors contribute to upregulation of beta-endorphin in inflamed skin tissues by electroacupuncture. Mol Pain 7:98, (Epub ahead of print) PMID: 22177137

Sztainberg Y, Kuperman Y, Issler O, Gil S, Vaughan J, Rivier J, Vale W, Chen A (2009) A novel corticotropin-releasing factor receptor splice variant exhibits dominant negative activity: a putative link to stress-induced heart disease. FASEB J 23:2186–2196

Tachibana T, Nawa T (2005) Immunohistochemical reactions of receptors to met-enkephalin, VIP, substance P, and CGRP located on Merkel cells in the rat sinus hair follicle. Arch Histol Cytol 68:383–391

Tachibana T, Taniguchi S, Furukawa F, Miwa S, Imamura S (1990) Serotonin metabolism in the arthus reaction. J Invest Dermatol 94:120–125

Tachibana T, Endoh M, Fujiwara N, Nawa T (2005) Receptors and transporter for serotonin in Merkel cell-nerve endings in the rat sinus hair follicle. An immunohistochemical study. Arch Histol Cytol 68:19–28

Tagen M, Stiles L, Kalogeromitros D, Gregoriou S, Kempuraj D, Makris M, Donelan J, Vasiadi M, Staurianeas NG, Theoharides TC (2007) Skin corticotropin-releasing hormone receptor expression in psoriasis. J Invest Dermatol 127:1789–1791

Takahashi H, Kinouchí M, Tamura T, Iizuka H (1996) Decreased beta 2-adrenergic receptor-mRNA and loricrin-mRNA, and increased involucrin-mRNA transcripts in psoriatic epidermis: analysis by reverse transcription-polymerase chain reaction. Br J Dermatol 134:1065–1069

Tan DX, Manchester LC, Burkhardt S, Sainz RM, Mayo JC, Kohen R, Shohami E, Huo YS, Hardeland R, Reiter RJ (2001) N1-acetyl-N2-formyl-5-methoxykynuramine, a biogenic amine and melatonin metabolite, functions as a potent antioxidant. FASEB J 15:2294–2296

Tan DX, Reiter RJ, Manchester LC, Yan MT, El Sawi M, Sainz RM, Mayo JC, Kohen R, Allegra M, Hardeland R (2002) Chemical and physical properties and potential mechanisms: melatonin as a broad spectrum antioxidant and free radical scavenger. Curr Top Med Chem 2:181–197

Taneda K, Tominaga M, Negi O, Tengara S, Kamo A, Ogawa H, Takamori K (2011) Evaluation of epidermal nerve density and opioid receptor levels in psoriatic itch. Br J Dermatol 165:277–284

Tang JY, Xiao TZ, Oda Y, Chang KS, Shpall E, Wu A, So PL, Hebert J, Bikle D, Epstein EH Jr (2011) Vitamin D3 inhibits hedgehog signaling and proliferation in murine basal cell carcinomas. Cancer Prev Res 4:744–751

Taves MD, Gomez-Sanchez CE, Soma KK (2011) Extra-adrenal glucocorticoids and mineralocorticoids: evidence for local synthesis, regulation, and function. Am J Physiol Endocrinol Metab 301:E11–E24

Tayebati SK, El-Assouad D, Ricci A, Amenta F (2002) Immunochemical and immunocytochemical characterization of cholinergic markers in human peripheral blood lymphocytes. J Neuroimmunol 132:147–155

Teichert AE, Elalieh H, Elias PM, Welsh J, Bikle DD (2011) Overexpression of hedgehog signaling is associated with epidermal tumor formation in vitamin D receptor-null mice. J Invest Dermatol 131:2289–2297

Telek A, Biro T, Bodo E, Toth BI, Borbiro I, Kunos G, Paus R (2007) Inhibition of human hair follicle growth by endo- and exocannabinoids. FASEB J 21:3534–3541

Theoharides TC, Cochrane DE (2004) Critical role of mast cells in inflammatory diseases and the effect of acute stress. J Neuroimmunol 146:1–12

Thiboutot D, Jabara S, Mcallister JM, Sivarajah A, Gilliland K, Cong Z, Clawson G (2003) Human skin is a steroidogenic tissue: steroidogenic enzymes and cofactors are expressed in epidermis, normal sebocytes, and an immortalized sebocyte cell line (SEB-1). J Invest Dermatol 120:905–914

Thorslund K, El-Nour H, Nordlind K (2009) The serotonin transporter protein is expressed in psoriasis, where it may play a role in regulating apoptosis. Arch Dermatol Res 301:449–457

Tian XQ, Holick MF (1999) A liposomal model that mimics the cutaneous production of vitamin D3. Studies of the mechanism of the membrane-enhanced thermal isomerization of previtamin D3 to vitamin D3. J Biol Chem 274:4174–4179

Tiede S, Bohm K, Meier N, Funk W, Paus R (2010) Endocrine controls of primary adult human stem cell biology: thyroid hormones stimulate keratin 15 expression, apoptosis, and differentiation in human hair follicle epithelial stem cells in situ and in vitro. Eur J Cell Biol 89:769–777

Tiganescu A, Walker EA, Hardy RS, Mayes AE, Stewart PM (2011) Localization, age- and site-dependent expression, and regulation of 11beta-hydroxysteroid dehydrogenase type 1 in skin. J Invest Dermatol 131:30–36

Tint GS, Irons M, Elias ER, Batta AK, Frieden R, Chen TS, Salen G (1994) Defective cholesterol biosynthesis associated with the Smith-Lemli-Opitz syndrome. N Engl J Med 330:107–113

Tobin DJ (2006) Biochemistry of human skin–our brain on the outside. Chem Soc Rev 35:52–67

Tobin DJ, Kauser S (2005a) Beta-endorphin: the forgotten hair follicle melanotropin. J Investig Dermatol Symp Proc 10:212–216

Tobin DJ, Kauser S (2005b) Hair melanocytes as neuro-endocrine sensors–pigments for our imagination. Mol Cell Endocrinol 243:1–11

Tominaga M, Ogawa H, Takamori K (2007) Possible roles of epidermal opioid systems in pruritus of atopic dermatitis. J Invest Dermatol 127:2228–2235

Tomlinson JW, Walker EA, Bujalska IJ, Draper N, Lavery GG, Cooper MS, Hewison M, Stewart PM (2004) 11beta-hydroxysteroid dehydrogenase type 1: a tissue-specific regulator of glucocorticoid response. Endocr Reviews 25:831–866

Toth BI, Dobrosi N, Dajnoki A, Czifra G, Olah A, Szollosi AG, Juhasz I, Sugawara K, Paus R, Biro T (2011) Endocannabinoids modulate human epidermal keratinocyte proliferation and survival via the sequential engagement of cannabinoid receptor-1 and transient receptor potential vanilloid-1. J Invest Dermatol 131:1095–1104

Tsang AH, Chung KK (2009) Oxidative and nitrosative stress in Parkinson's disease. Biochim Biophys Acta 1792:643–650

Tsuji H, Okamoto K, Matsuzaka Y, Iizuka H, Tamiya G, Inoko H (2003) SLURP-2, a novel member of the human Ly-6 superfamily that is up-regulated in psoriasis vulgaris. Genomics 81:26–33

Tuckey RC (2005) Progesterone synthesis by the human placenta. Placenta 26:273 281

Tuckey RC, Nguyen MN, Chen J, Slominski AT, Baldisseri DM, Tieu EW, Zjawiony JK, Li W (2011) Human cytochrome P450scc (CYP11A1) catalyses epoxide formation with ergosterol. Drug Metab Dispos 40(3):436–444, (Epub ahead of print) PMID: 22106170

Vacchio MS, Papadopoulos V, Ashwell JD (1994) Steroid production in the thymus: implications for thymocyte selection. J Exp Med 179:1835–1846

Vale W, Spiess J, Rivier C, Rivier J (1981) Characterization of a 41-residue ovine hypothalamic peptide that stimulates secretion of corticotropin and beta-endorphin. Science 213:1394–1397

Valencia A, Kochevar IE (2006) Ultraviolet al. induces apoptosis via reactive oxygen species in a model for Smith-Lemli-Opitz syndrome. Free Radic Biol Med 40:641–650

van Beek N, Bodo E, Kromminga A, Gaspar E, Meyer K, Zmijewski MA, Slominski A, Wenzel BE, Paus R (2008) Thyroid hormones directly alter human hair follicle functions: anagen prolongation and stimulation of both hair matrix keratinocyte proliferation and hair pigmentation. J Clin Endocrinol Metab 93:4381–4388

Vaudry H, Chartrel N, Desrues L, Galas L, Kikuyama S, Mor A, Nicolas P, Tonon MC (1999) The pituitary-skin connection in amphibians. Reciprocal regulation of melanotrope cells and dermal melanocytes. Ann N Y Acad Sci 885:41–56

Vukelic S, Stojadinovic O, Pastar I, Rabach M, Krzyzanowska A, Lebrun E, Davis SC, Resnik S, Brem H, Tomic-Canic M (2011) Cortisol synthesis in epidermis is induced by IL-1 and tissue injury. J Biol Chem 286:10265–10275

Waller LA, Sarkka A, Olsbo V, Myllymaki M, Panoutsopoulou IG, Kennedy WR, Wendelschafer-Crabb G (2011) Second-order spatial analysis of epidermal nerve fibers. Stat Med 30:2827–2841

Wallstrom M, Sand L, Nilsson F, Hirsch JM (1999) The long-term effect of nicotine on the oral mucosa. Addiction 94:417–423

Walterscheid JP, Nghiem DX, Kazimi N, Nutt LK, Mcconkey DJ, Norval M, Ullrich SE (2006) Cis-urocanic acid, a sunlight-induced immunosuppressive factor, activates immune suppression via the 5-HT2A receptor. Proc Natl Acad Sci USA 103:17420–17425

Wasserman D, Sokolowski M, Rozanov V, Wasserman J (2008) The CRHR1 gene: a marker for suicidality in depressed males exposed to low stress. Genes Brain Behav 7:14–19

Watson S (1994) Dopamine receptors. In: Watson S, Arkinstall S (eds) The G-protein linked receptor fact book. Academic Press, London, UK

Weedon D (2010) Weedon's skin pathology. Churchill Livingstone—Elsevier, Oxford, UK

Welch DA, Samuel WM, Hudson RJ (1990) Bioenergetic consequences of alopecia induced by *Dermacentor albipictus* (Acari: Ixodidae) on moose. J Med Entomol 27:656–660

Wessler I, Reinheimer T, Kilbinger H, Bittinger F, Kirkpatrick CJ, Saloga J, Knop J (2003) Increased acetylcholine levels in skin biopsies of patients with atopic dermatitis. Life Sci 72:2169–2172

Wiesenberg I, Missbach M, Carlberg C (1998) The potential role of the transcription factor RZR/ROR as a mediator of nuclear melatonin signaling. Restor Neurol Neurosci 12:143–150

Wiesner B, Roloff B, Fechner K, Slominski A (2003) Intracellular calcium measurements of single human skin cells after stimulation with corticotropin-releasing factor and urocortin using confocal laser scanning microscopy. J Cell Sci 116:1261–1268

Wietfeld D, Heinrich N, Furkert J, Fechner K, Beyermann M, Bienert M, Berger H (2004) Regulation of the coupling to different G proteins of rat corticotropin-releasing factor receptor type 1 in human embryonic kidney 293 cells. J Biol Chem 279:38386–38394

Wintzen M, Gilchrest BA (1996) Proopiomelanocortin, its derived peptides, and the skin. J Invest Dermatol 106:3–10

Wintzen M, Ostijn DM, Polderman MC, Le Cessie S, Burbach JP, Vermeer BJ (2001) Total body exposure to ultraviolet radiation does not influence plasma levels of immunoreactive beta-endorphin in man. Photodermatol Photoimmunol Photomed 17:256–260

Wu J, Lukas RJ (2011) Naturally-expressed nicotinic acetylcholine receptor subtypes. Biochem Pharmacol 82:800–807

Yang G, Zhang G, Pittelkow MR, Ramoni M, Tsao H (2006) Expression profiling of UVB response in melanocytes identifies a set of p53-target genes. J Invest Dermatol 126:2490–2506

Yen PM (2001) Physiological and molecular basis of thyroid hormone action. Physiol Rev 81:1097–1142

Yoshida M, Takahashi Y, Inoue S (2000) Histamine induces melanogenesis and morphologic changes by protein kinase A activation via H2 receptors in human normal melanocytes. J Invest Dermatol 114:334–342

Yosipovitch G (2010) Chronic pruritus: a paraneoplastic sign. Dermatol Ther 23:590–596

Young SF, Griffante C, Aguilera G (2007) Dimerization between vasopressin V1b and corticotropin releasing hormone type 1 receptors. Cell Mol Neurobiol 27:439–461

Yu HS, Reiter RJ (1993) Melatonin biosynthesis, physiological effects, and clinical implications. CRC Press, Boca Raton

Yuasa T, Ono M, Watanabe T, Takai T (2001) Lyn is essential for fcgamma receptor III-mediated systemic anaphylaxis but not for the Arthus reaction. J Exp Med 193:563–572

Zagon IS, Wu Y, Mclaughlin PJ (1996) The opioid growth factor, [Met5]-enkephalin, and the zeta opioid receptor are present in human and mouse skin and tonically act to inhibit DNA synthesis in the epidermis. J Invest Dermatol 106:490–497

Zbytek B, Slominski AT (2007) CRH mediates inflammation induced by lipopolysaccharide in human adult epidermal keratinocytes. J Invest Dermatol 127:730–732

Zbytek B, Mysliwski A, Slominski A, Wortsman J, Wei ET, Mysliwska J (2002) Corticotropin-releasing hormone affects cytokine production in human HaCaT keratinocytes. Life Sci 70:1013–1021

Zbytek B, Pfeffer LM, Slominski AT (2004) Corticotropin releasing hormone stimulates NF-kappa B in human epidermal keratinocytes. J Endocrinol 181:R1–R7

Zbytek B, Pikula M, Slominski RM, Mysliwski A, Wei E, Wortsman J, Slominski AT (2005) Corticotropin-releasing hormone triggers differentiation in HaCaT keratinocytes. Br J Dermatol 152:474–480

Zbytek B, Pfeffer LM, Slominski AT (2006a) CRH inhibits NF-kappa B signaling in human melanocytes. Peptides 27:3276–3283

Zbytek B, Wortsman J, Slominski A (2006b) Characterization of a ultraviolet B-induced corticotropin-releasing hormone-proopiomelanocortin system in human melanocytes. Mol Endocrinol 20:2539–2547

Zbytek B, Janjetovic Z, Tuckey RC, Zmijewski MA, Sweatman TW, Jones E, Nguyen MN, Slominski AT (2008) 20-Hydroxyvitamin D3, a product of vitamin D3 hydroxylation by cytochrome P450scc, stimulates keratinocyte differentiation. J Invest Dermatol 128:2271–2280

Zhang X, Beaulieu JM, Sotnikova TD, Gainetdinov RR, Caron MG (2004) Tryptophan hydroxylase-2 controls brain serotonin synthesis. Science 305:217

Zhang M, Thurmond RL, Dunford PJ (2007) The histamine H(4) receptor: a novel modulator of inflammatory and immune disorders. Pharmacol Ther 113:594–606

Zhao ZG, Li YY, Yang J, Li HJ, Zhao H (2010) Expression of cannabinoid receptor 2 in squamous cell carcinoma. Nan Fang Yi Ke Da Xue Xue Bao 30:593–595

Zheng D, Bode AM, Zhao Q, Cho YY, Zhu F, Ma WY, Dong Z (2008) The cannabinoid receptors are required for ultraviolet-induced inflammation and skin cancer development. Cancer Res 68:3992–3998

Zmijewski MA, Slominski AT (2009a) CRF1 receptor splicing in epidermal keratinocytes: potential biological role and environmental regulations. J Cell Physiol 218:593–602

Zmijewski MA, Slominski AT (2009b) Modulation of corticotropin releasing factor (CRF) signaling through receptor splicing in mouse pituitary cell line AtT-20—emerging role of soluble isoforms. J Physiol Pharmacol 60(Suppl 4):39–46

Zmijewski MA, Slominski AT (2010) Emerging role of alternative splicing of CRF1 receptor in CRF signaling. Acta Biochim Pol 57:1–13

Zmijewski MA, Slominski AT (2011) Neuroendocrinology of the skin: an overview and selective analysis. Dermatoendocrinology 3:3–10

Zmijewski MA, Sharma RK, Slominski AT (2007) Expression of molecular equivalent of hypo-thalamic-pituitary-adrenal axis in adult retinal pigment epithelium. J Endocrinol 193:157–169

Zmijewski MA, Li W, Zjawiony JK, Sweatman TW, Chen J, Miller DD, Slominski AT (2009a) Photo-conversion of two epimers (20R and 20S) of pregna-5,7-diene-3beta, 17alpha, 20-triol and their bioactivity in melanoma cells. Steroids 74:218–228

Zmijewski MA, Sweatman TW, Slominski AT (2009b) The melatonin-producing system is fully functional in retinal pigment epithelium (ARPE-19). Mol Cell Endocrinol 307:211–216

Zmijewski MA, Li W, Chen J, Kim TK, Zjawiony JK, Sweatman TW, Miller DD, Slominski AT (2011) Synthesis and photochemical transformation of 3beta,21-dihydroxypregna-5,7-dien-20-one to novel secosteroids that show anti-melanoma activity. Steroids 76:193–203

Zouboulis CC (2004) Acne and sebaceous gland function. Clin Dermatol 22:360–366

Zouboulis CC, Degitz K (2004) Androgen action on human skin—from basic research to clinical significance. Exp Dermatol 13(Suppl 4):5–10

Zouboulis CC, Seltmann H, Hiroi N, Chen W, Young M, Oeff M, Scherbaum WA, Orfanos CE, Mccann SM, Bornstein SR (2002) Corticotropin-releasing hormone: an autocrine hormone that promotes lipogenesis in human sebocytes. Proc Natl Acad Sci USA 99:7148–7153

Zouboulis CC, Chen WC, Thornton MJ, Qin K, Rosenfield R (2007) Sexual hormones in human skin. Horm Metab Res 39:85–95

Zouboulis CC, Baron JM, Bohm M, Kippenberger S, Kurzen H, Reichrath J, Thielitz A (2008) Frontiers in sebaceous gland biology and pathology. Exp Dermatol 17:542–551